Ji Lianghai

Design and Application of D2D and V2X Communications in the 5G Radio Access Network

Logos Verlag Berlin

λογος

Bibliographic information published by Die Deutsche Bibliothek

Die Deutsche Bibliothek lists this publication in the Deutsche Nationalbibliografie; detailed bibliographic data is available in the Internet at http://dnb.ddb.de.

ISBN 978-3-8325-5030-1

Logos Verlag Berlin GmbH
Comeniushof, Gubener Str. 47,
D-10243 Berlin
Germany

Tel.: +49 (0)30 / 42 85 10 90
Fax: +49 (0)30 / 42 85 10 92
http://www.logos-verlag.de

Design and Application of D2D and V2X Communications in the 5G Radio Access Network

Design und Anwendung von D2D- und V2X-Kommunikation im 5G-Funkzugangsnetz

von

M.Sc. Lianghai Ji

geboren in Shandong

Vom Fachbereich Elektrotechnik und Informationstechnik
der Technischen Universität Kaiserslautern
zur Verleihung des akademischen Grades
Doktor der Ingenieurswissenschaften (Dr.-Ing.)
genehmigte Dissertation

D 386

Datum der mündlichen Prüfung: 24.10.2019

Dekan des Fachbereiches: Prof. Dr.-Ing. Ralph Urbansky
Prüfungsvorsitzender: Prof. Dr.-Ing. habil. Norbert Wehn

1. Berichterstatter: Prof. Dr.-Ing. Hans D. Schotten
2. Berichterstatter: Prof. Dr.-Ing. habil. Dr.-Ing. E.h. Paul Walter Baier

Dedicated to my parents Ji Lin and Peng Qin, and also my beloved Siya

Zusammenfassung

Es wird erwartet, dass die Nachfrage nach Proximity-Services (ProSe) in Mobilfunksystemen über das Jahr 2020 hinaus in verschiedenen Anwendungsbereichen stark ansteigen wird, z. B. im Bereich sozialer Apps und im Bereich öffentlicher Verkehrsmittel. ProSe bezeichnet Dienste zwischen Mobilfunkgeräten, welche sich in geografischer Nähe zueinander befinden und nach dem Durchlaufen einer sogenannten Entdeckungsphase lokal Informationen untereinander austauschen. Klassischerweise erfolgt in zellularen Netzwerken, z.B. 3G oder 4G, der Austausch von Informationen über die Basisstation zwischen den Endgeräten. Hierfür werden entsprechende Ressourcen für den Up- und den Downlink bereitgestellt. Im Vergleich dazu wird bei ProSe eine direkte Kommunikation zwischen zwei geografisch benachbarten Endgeräten ermöglicht, ohne die Informationen über die Netzinfrastruktur zu senden. Die Unterstützung von ProSe gilt als einer des wichtigsten technischen Wegbereiter für Mobilfunksysteme der nächsten Generation (5G). Der direkte Kommunikationsmodus, welcher auch als Device-to-Device (D2D)-Kommunikation bezeichnet wird, bietet gegenüber dem klassischen zellularen Modus eine Vielzahl an Vorteilen. Hierzu zählen insbesondere eine verbesserte spektrale Effizienz, eine bessere Leistungseffizienz sowie eine reduzierte Übertragungslatenz und eine Vergrößerung der Funkabdeckung. Innerhalb dieser Arbeit wird insbesondere untersucht, inwiefern durch die Anwendung von D2D-Kommunikation eine Steigerung der Leistungsfähigkeit und Effizienz von Mobilfunksystemen für die verschiedene Arten von ProSe erzielt werden kann.

Eine der Anforderungen an Mobilfunksystemen der fünften Generation ist, dass das übertragbare Datenvolumen im Vergleich zu den System der vierten Generation noch einmal deutlich gesteigert werder kann. Um diese Anforderung zu erfüllen, bietet die D2D-Kommunikation eine Möglichkeit zur Entlastung der zellularen Verbindungen, indem ein Teil des Datenverkehrs in einen lokalen Content-Sharing-Prozess ausgelagert wird. Dabei wird insbesondere der netzwerkgestützte Operationsmodus der D2D-Kommunikation, welcher durch das primäre zellulare System, z.B. Long-Term-Evolution (LTE), gesteuert wird und dessen Funkressourcen wiederverwendet, als eine Option zur Erhöhung der Systemkapazität betrachtet. Durch die Wiederverwendung der Funkressourcen der zellularen Verbindungen durch D2D-Verbindungen tritt jedoch das Problem der gegenseitigen Interferenz auf. Daher ist es wichtig, effiziente Mechanismen einzuführen, um Interferenzen kontrollieren und minimieren zu können. Im Rahmen dieser Arbeit werden dazu netzwerkgesteuerte Radio Resource Management (RRM) Algorithmen mit geringer Berechnungskomplexität entwickelt, um die Ressourcenwiederverwendung zu optimieren. Diese werden hinsichtlich der relevanten Key Performance Indikatoren, z.B. Anzahl der unterstützten D2D Verbindungen, Systemkapazität und Benutzerzufriedenheit, untersucht. Um eine optimale Lösung für die entwickelten RRM-Algorithmen zu erzielen, werden zusätzlich Kontextinformationen berücksichtigt. Dazu werden in dieser Arbeit weiterhin Signalisierungsschemata mit geringem Signalisierungsaufwand entworfen. Mit Hilfe von numerischen Ergebnissen, welche aus Simulationen auf Systemebene (System Level Simulation) gewonnen wurden, wird die erzielbare Leistungssteigerung evaluiert.

Neben dem Ziel der Verbesserung der Servicequalität für Breitbanddienste der Klasse „Enhanced Mobile Broadband“ (eMBB) liegt der Fokus von 5G auch auf anderen neu entstehenden Diensten. Diese sind insbesondere Dienste im Bereich „Internet-of-Things“ (IoT), welche auch als „Massive Machine Type Communication“ (mMTC) bezeichnet werden und beispielsweise das Potenzial haben, neue Märkte für zukünftige Mobilfunknetze zu erschließen. Die Anforderungen der aufkommenden mMTC-Anwendungen können jedoch von den derzeitigen Mobilfunksystemen, z. B. LTE, nicht erfüllt werden, da diese in erster Linie für eMMB Dienste ausgelegt sind. Bisher wurde sowohl im wissenschaftlichen als auch im kommerziellen Bereich in Studien untersucht, inwiefern die Unterstützung für mMTC-Dienste durch eine Verbesserung und Anpassung der Mobilfunksysteme der vierten Generation ermöglicht werden kann. Die Herausforderungen dabei sind insbesondere, die Verfügbarkeit der Dienste zu verbessern und gleichzeitig den Energieverbrauch von mMTC-Endgeräten zu reduzieren. In dieser Arbeit wird dazu ein netzwerkgesteuertes Sidelink-Kommunikationsschema erarbeitet, welches unter Heranziehung von verschiedenen Kontextinformationen die Effizienz von mMTC-unterstützen Mobilfunknetzen erhöht. Mit Hilfe der Kontextinformationen kann eine intelligente Konfiguration sowohl für die direkte als auch für die zellulare Verbindung ermöglicht werden. Zur Akquisition der dazu relevanten Kontextinformationen wird eine entsprechende aufwandsarme Signalisierung entworfen. Dazu wird im Rahmen dieser Arbeit weiterhin ein Simulator implementiert, um die vorgeschlagenen Konzepte sowohl in Szenarien mit eine einziger Mobilfunkzelle als auch mit mehreren Zellen zu evaluieren. Die Simulationsergebnisse zeigen dabei, dass durch die vorgeschlagenen Lösungsansätze gleichzeitig sowohl die Serviceverfügbarkeit als auch die Energieeffizienz von mMTC-Endgeräten verbessert werden kann.

Eine weitere Diensteklasse in zukünftigen zellularen Systemen sind Dienste, welche den Informationsaustausch zwischen verschiedenen Verkehrsteilnehmern ermöglichen. Diese werden auch als „Vehicle-to-Everything" (V2X)-Kommunikation bezeichnet und haben als ein Ziel, den Automatisierungsgrad in Fahrzeugen durch entsprechende Anwendungen zu erhöhen, beispielsweise um Unfälle zu vermeiden oder eine Verbesserung der Verkehrseffizienz zu erzielen. Um diese Anwendungen zu unterstützen, sind eine extrem hohe Zuverlässigkeit und eine niedrige Übertragungslatenz erforderlich, welche für die Kommunikationstechnologien der vorherigen Generation (z. B. LTE) technisch sehr herausfordernd sind. Im Rahmen dieser Arbeit werden zunächst die beiden Funkschnittstellen LTE-Uu für zellulare Verbindungen mit involvierter Netzwerkinfrastruktur zur Übertragung des V2X-Datenverkehrs und PC5 für Übertragung des V2X-Datenverkehrs über direkte Verbindungen zwischen Endgeräten untersucht, welche vom 3rd Generation Partnership Project (3GPP) als mögliche Technologien vorgeschlagen wurden, um V2X-Kommunikation zu ermöglichen. Eine weitere Anforderung an diese Technologien ist eine große Reichweite für V2X-Kommunikation aufgrund der hohen Geschwindigkeiten von Fahrzeugen, bspw. auf Autobahnen. Dies kann mit einer einzigen Paketübertragung (Single Hop) eines Senders über die direkte Funkschnittstelle nicht erreicht werden. Daher wird in dieser Arbeit ein kontextsensitives V2X-Kommunikationsschema eingeführt, bei welchem einige Fahrzeuge ausgewählt werden, um als Relais zu fungieren und ihre empfangenen Pakete erneut an andere Fahrzeuge zu senden. Dies ermöglicht auch eine Datenübertragung, obwohl sich die Empfänger möglicherweise außerhalb der Reichweite des ursprünglichen Senders befinden (Two Hop Übertragung). Mit diesem Ansatz können sowohl die Reichweite als auch die Zuverlässigkeit der entsprechenden Paketübertragung im Rahmen der V2X-Kommunikation erhöht werden. Die jeweiligen Funkressourcen dieser beiden Übertragungen werden dabei basierend auf der Lösung des zugehörigen Optimierungsproblems zugeordnet. Eine weitere Anforderung von V2X-Services ist die Gewährleistung einer extrem hohen Zuverlässigkeit, welche in einigen Szenarien nicht mit einer einzigen Übertragungstechnologie, beispielsweise LTE-Uu oder PC5, erfüllt werden kann. Daher wurde in diesem Zusammenhang untersucht, inwiefern diese Anforderung durch das gleichzeitige Nutzen zweier oder mehrerer Funktechnologien, was auch als „Multi Radio Access Technology“ (Multi-RAT) bezeichnet wird, erfüllt werden kann. Um das Multi-RAT Konzept auf effiziente Weise anzuwenden, werden verschiedene Multi-RAT-Verfahren entworfen und verglichen. Dabei wird auch das klassische Uplink-Konzept für den Fall erweitert, dass eine Basisstation V2X-Datenpakete sowohl von der LTE-Uu- als auch von der PC5-Schnittstelle empfangen kann. Für den Downlink werden zur Verbesserung der Übertragungszuverlässigkeit verschiedene Multicast-Broadcast-Single-Frequency-Network-(MBSFN)-Area-Mapping-Ansätze entwickelt.

Schlagwörter - 5G-Funkzugangsnetz, Proximity-basierte Dienste, Geräte-zu-Gerät-Kommunikation, massive Maschinentyp-Kommunikation, Vehicle-to-Everything-Kommunikation, Bewertungsmethodik

Abstract

It is commonly expected that there will be an evolving demand for proximity-based services (ProSe) in different application areas for the time beyond 2020, e.g. social apps and public transportation. ProSe denotes services requiring a direct discovery and communication among mobile devices that are geographically located near each other. In this sense, a local information exchange is required to support the ProSe. In legacy cellular networks, e.g. 3G and 4G, this procedure is realized by transmitting data traffic through a base station via uplink and downlink. In contrast to that, a direct communication mode not transmitting data traffic through any network infrastructure is referred to as sidelink communication. It has been considered as one of the key technical enablers for the next generation of mobile communications systems (5G), especially from the perspective of supporting ProSe. With the direct communication mode, a direct device-to-device (D2D) communication can be applied with certain benefits, e.g. an improved spectral efficiency, a better power efficiency, a reduced latency, and an extended coverage area.

In this thesis, the author investigates and develops how to efficiently apply the direct communication mode to improve the performances of different 5G service types, i.e. enhanced Mobile Broadband (eMBB) service, massive Machine Type Communication (mMTC), and Vehicle-to-Everything (V2X) communication. In particular, context-aware algorithms are developed to optimize network configurations. Moreover, in order to support the proposed direct communication technologies, corresponding signaling schemes are designed. Last but not least, simulations are conducted to evaluate the proposed schemes quantitatively.

Keywords - 5G Radio Access Network, Proximity-based Services, Evaluation Methodology, Device-to-Device Communication, massive Machine Type Communications, Vehicle-to-Everything Communication

Contents

List of Figures

List of Tables

List of Acronyms

2D	two-dimensional
3D	three-dimensional
3G	third generation of mobile communication system
3GPP	3rd Generation Partnership Project
4G	fourth generation of mobile communication system
5G	next generation of mobile communication system
ACK	acknowledgment
AM	acknowledged mode
AoA	angle of arrival
AoD	angle of departure
BLER	block error rate
BM-SC	Broadcast Multicast - Service Center
BS	base station
C-ITS	Cooperative Intelligent Transport Systems
CAM	Cooperative Awareness Message
CDF	cumulative distribution function
CP	control-plane
CPE	collective perception of the environment
CRC	cyclic redundancy check
CSI	channel state information
CSMA/CA	carrier-sense multiple access with collision avoidance
CSRS	cell-specific reference signals
D2D	Device-to-Device
DCI	downlink control information
DL	downlink
DMRS	demodulation reference signals
DRX	discontinuous reception
E2E	end-to-end

EESM	exponential effective SINR metric
eMBB	enhanced Mobile Broadband
ETSI	European Telecommunication Standards Institute
FDD	frequency-division duplex
GPS	Global Positioning System
HARQ	hybrid automatic repeat request
I2I	indoor-to-indoor
ID	identity
IEEE	Institute of Electrical and Electronics Engineers
IMT-2020	IMT for 2020 and beyond
IMT-Advanced	International Mobile Telecommunications-Advanced
ISD	inter-site distance
ISI	inter-symbol interference
ITU-R	ITU Radiocommunication Sector
IVD	inter-vehicle distance
KPI	key performance indicator
L2S	link-to-system
LOS	line-of-sight
LTE	Long Term Evolution
LTE-M	Long Term Evolution for Machines
LTE-Uu	definition from 3GPP for uplink and downlink
MAC	medium access control
MBB	mobile broadband
MBMS	multimedia broadcast multicast services
MBMS-GW	MBMS gateway
MBSFN	multicast-broadcast single-frequency network
MCS	modulation and coding scheme
MEC	mobile edge computing
MIB	master information block
MIESM	mutual information effective SINR metric

MIMO	multiple-input and multiple-output
mMTC	massive Machine Type Communications
MNO	mobile network operator
MRC	maximal ratio combining
MTD	machine-type device
multi-RAT	multi-radio access technologies
NACK	non-acknowledgment
NLOS	non-line-of-sight
O2I	outdoor-to-indoor
O2O	outdoor-to-outdoor
OTT	one trip time
P2MP	point-to-multi-point
P2P	point-to-point
PA	power amplifier
PC5	definition from 3GPP for sidelink
ProSe	proximity-based services
PRR	packet reception ratio
PUSCH	physical uplink shared channel
QoE	quality of experience
QoS	quality of service
RA	random access
RAN	radio access network
RAT	radio access technology
RB	resource block
RLC	radio link control
RMa	rural macro
RRC	Radio Resource Control
RRC_CONNECTED	3GPP definition for the user in the connected state
RRC_IDLE	3GPP definition for the user in the idle state
RRM	radio resource management

RSU	Roadside Unit
Rx	receiver
S-RSRP	sidelink reference signal received power
S-SIB	sidelink system information block
SIB	system information block
single-RAT	single-radio access technology
SINR	signal-to-interference-plus-noise ratio
SL	sidelink
SLA	service level agreement
SPS	semi-persistent scheduling
SR	scheduling request
TDD	time-division duplex
TM	transmission mode
TMS	transmission mode selection
TTI	transmission time interval
Tx	transmitter
UE	user equipment
UL	uplink
UM	unacknowledged mode
UMa	urban macro
UMi	urban micro
UP	user-plane
URLLC	Ultra-Reliable Low Latency Communications
V2I	Vehicle-to-Infrastructure
V2N	Vehicle-to-Network
V2P	Vehicle-to-Pedestrian
V2V	Vehicle-to-Vehicle
V2X	Vehicle-to-Everything

List of Frequently Used Symbols

γ	function to calculate the threshold value
γ_{cell}	the deterioration threshold value for the cellular link
ζ	pathloss exponent
σ	shadow fading standard deviation
σ_{n}^2	noise power
λ	wavelength
φ_i	reference angle for the mMTC device in the i-th cluster
$\varphi_i^{\mathrm{start}}$	reference angle where the i-th cluster starts
φ_i^{end}	reference angle where the i-th cluster ends
a_m	the mth element in vector A
A	satisfaction degree vector
A_{sector}	the area of one cluster
A_{p}	a constant component in pathloss
B_{required}	number of required RBs
$BC_{(i,j)}(t)$	the real-time battery capacity of user-j in the cluster i at the time instance t
BL_{cell}	battery life of the sensor by using the cellular uplink
$BL_{\mathrm{threshold}}$	battery life requirement
BW	bandwidth
BW_1	bandwidth allocated to the first hop
BW_2	bandwidth allocated to the second hop
c	propagation spead of light
$c(m,n)$	achievable capacity if the nth and the mth links use the same resource

$c^{\mathrm{CD}}(m,n)$	achievable capacity if the resource of the nth cellular UE is reused by the mth D2D pair
C	system capacity
$\boldsymbol{C}$	capacity matrix if two D2D links can reuse the same resource
$\boldsymbol{C}(m,:)$	the mth row of the matrix $\boldsymbol{C}$
$\boldsymbol{C}^{\mathrm{CD}}$	capacity matrix if a D2D link is only allowed to reuse the resource of a cellular link
$\boldsymbol{C}^{\mathrm{CD}}(m,:)$	the mth row of the matrix $\boldsymbol{C}^{\mathrm{CD}}$
$CellularNum$	ID of the combined cellular users
d	signal propagation distance in 2D
d_{3D}	signal propagation distance in 3D
d_{BP}	distanct at the breaking point
d_m	distance between the two ends of the mth D2D pair
$d_{(m,n)}$	distance from the nth cellular UE to the Rx of the mth D2D pair
d_{fl}	height of one floor
d_{p}	penetration distance
d_{r}	distances from the Rx to the intersection point
d_{t}	distances from the Tx to the intersection point
D_{rx}	number of DRX cycles per day
$D2DpairNum$	ID of the combined D2D pairs
E	price vector
E_n	the nth element in vector E
$EC_{(i,j)}$	the energy consumption of the user-j in the cluster i by using the cellular uplink for a unit time

f_{c}	carrier frequency
$f(m,n)$	feasibility function if the nth cellular link and the mth D2D link use the same resource
$\tilde{f}(m,n)$	distance-based feasibility function if the nth cellular link and the mth D2D link use the same resource
$\boldsymbol{F}$	feasibility matrix
$\boldsymbol{F}'$	manualy constructed matrix
$FeasibleLinkIDs$	IDs of links using the same RB
h	average building height
h_{BS}	actual antenna height at the BS
h_{UT}	actual antenna height at the UE
h'_{BS}	effective antenna height at the BS
h'_{UT}	effective antenna height at the UE
h'_{rx}	effective antenna height at the Rx
h'_{tx}	effective antenna height at the Tx
$h_{(\mathrm{BS},m)}$	channel gain between the mth D2D Tx and the BS
h_m^{D2D}	channel gain between the two transmission ends of the mth D2D pair
$h_{(m,n)}$	channel gain from the nth cellular UE to the mth D2D Rx
h_n^{cell}	channel gain of the nth cellular link
$h_{(n,\hat{n})}$	channel gain from the $\hat{n}^{\mathrm{th}}$ D2D Tx to n^{th} D2D Rx
$I(i,j)$	the interference power when the i-th Rx receives the packet from the j-th Tx
$InfeasibleLinkIDs$	IDs of links which cannot reuse any RB
K	number of clusters
n_{f}	number of penetrated floors

N_{v}	the number of vehicles
N_{cell}	number of cellular links
N_{p}	number of penetrated internal walls
N_{D2D}	number of D2D links
$NormC(m,n)$	the element in the mth row and the nth column in the matrix $\boldsymbol{NormC}$
$\boldsymbol{NormC}(m,:)$	the mth row in the matrix $\boldsymbol{NormC}$
$\boldsymbol{NormC}$	matrix used in Algorithm 3
$NumOfLinks$	total number of links
NZ_{L}	number of zones in length
NZ_{W}	number of zones in width
O	asymptotic notation
pr_m	the mth element in vector PR
P	permutation operation
P_{LOS}	LOS probability
$P^{\mathrm{D}}_{\mathrm{clock}}$	device power consumption to obtain synchronization
$P^{\mathrm{D}}_{\mathrm{cp}}$	device power consumption during the CP establishment procedure
$P^{\mathrm{D}}_{\mathrm{paging}}$	device power consumption to receive paging command
$P^{\mathrm{D}}_{\mathrm{rx}}$	device power consumption to receive packets from remote sensors
$P^{\mathrm{D}}_{\mathrm{sleep}}$	device power consumption in sleeping mode
P_{tx}	transmit power of a device
$P^{\mathrm{cell}}_{\mathrm{tx}}$	transmit power of the cellular uplink
$P^{\mathrm{D}}_{\mathrm{tx}}$	device power consumption during transmission
$P^{\mathrm{D2D}}_{\mathrm{tx}}$	transmit power of the D2D link
P_{max}	maximal transmit power

P_{rx}	received signal power
$P_{\mathrm{rx}}^{\mathrm{target}}$	target received signal power
PL	pathloss
PL_{bpl}	building penetration loss
PL_{ew}	external wall penetration loss
PL_{fs}	free space pathloss
PL_{msd}	multiple screen diffraction loss
PL_{rts}	diffraction loss from rooftop to the street
PL_{LOS}	pathloss for LOS propagation
PL_{NLOS}	pathloss for NLOS propagation
$Prob$	probability
$Prob^{\mathrm{threshold}}$	probability threshold
PR	priority vector
r_m	data rate requirement of the mth link
R	data rate requirement vector
R_i	the distance from the BS to the mMTC device in the i-th cluster
R_{in}	the radius of the area without applying sidelink communication for mMTC services
R_i^{end}	the distance from the BS to the outer circle of the i-th cluster
R_i^{start}	the distance from the BS to the inner circle of the i-th cluster
R_{P}	packet transmission frequency of V2X communication
S	packet size of V2X communication
SE	spectral efficiency
$SINR_{\mathrm{target}}^{\mathrm{cell}}$	target SINR requirement to set up a cellular link

$SINR_{\text{target}}^{\text{D2D}}$	target SINR requirement to set up a D2D link
$SINR_{(m,n)}^{\text{cell}}$	SINR of the nth cellular uplink if its resource is reused by the mth D2D link
$SINR_{(m,n)}^{\text{D2D}}$	SINR of the mth D2D link if it reuses the resource of the nth cellular uplink
$SINR_n$	SINR value of the nth cellular link when its resource is not reused by any D2D link
$SINR_{(n,\hat{n})}^{\text{D2D}}$	SINR value of the nth D2D link if it uses the resource of the $\hat{n}^{\text{th}}$ D2D link
$SNR^{(i,j)}$	SNR of the cellular uplink for the user-j in the cluster i
$SNR_{\text{threshold}}$	SNR threshold value for the cellular uplink
t_{A}	the amount of time that the device has been served
$t_{\text{F}}^{\text{cell}}$	the amount of time that the device can be served by the cellular uplink
t_{DRX}	length of the DRX cycle
T_{TMS}	periodicity of the TM update procedure
W	street width
W_{i}	penetration loss of the internal walls
W_{r}	width of the street where the Rx locates
W_{t}	width of the street where the Tx locates
$x(m,n)$	resource allocation variable to show the number of RBs allocated to the nth and the mth links
$\boldsymbol{X}$	resource allocation matrix if two D2D links can reuse the same resource

$x^{\mathrm{CD}}(m,n)$	resource allocation variable to show if the mth D2D link and the nth cellular uplink use the same resource
$\boldsymbol{X}^{\mathrm{CD}}$	resource allocation matrix if a D2D link is only allowed to reuse the resource of a cellular link
X_{t}	distance from the Tx to the building wall
Z_{L}	length of a zone
Z_{W}	width of a zone

Chapter 1
Introduction

1.1 Motivation

In the past decades, wireless communication systems [Lin06, TV05, Wal99, Stu96] have made a rapid progress and are replacing the use of wired networks. Moreover, the services supported by wireless networks tend to become more and more diverse, e.g. covering both human-driven and machine-type services. The next generation of mobile communication system (5G) aims at providing enhanced human-dominated wireless communications as well an all-connected world of humans and objects [OMM16]. For example, 5G should enhance its mobile broadband (MBB) services with a better quality of service (QoS) compared with the legacy cellular systems [OBB+14], e.g. the third generation of mobile communication system (3G) and the fourth generation of mobile communication system (4G) [DPSB10, SBT11, HT07, HMCK05]. Moreover, there is a broad consensus that 5G will not only be a simple evolution of the 4G network with a higher spectral efficiency and a higher peak throughput, but also targets new services and business models [EMM+18]. As a typical 5G use case, in massive Machine Type Communications (mMTC) scenarios, a large amount, e.g. 300,000 machine-type devices (MTDs) within one cell [OBB+14] are expected to be connected with the network. For instance, in order to track objects in logistics, tags are attached to the objects and they need to be equipped with the capability for transmitting their location information to the network. Another typical example refers to agriculture applications, where low-cost sensors are distributed in rural areas to monitor the environment. In addition to the MBB and mMTC use cases, 5G will also provide support for the Ultra-Reliable Low Latency Communications (URLLC), where a low latency. e.g. 5 ms [OBB+14] and high reliability, e.g. 99.999% [OBB+14], are desired. One exemplary application which falls into this category is the automatic driving, which aims at better traffic safety and efficiency.

As a key technology component in 5G, the direct Device-to-Device (D2D) communication can facilitate a direct communication among nearby devices without transmitting data traffic through a network infrastructure. In this thesis, the author will focus on its design and application in the radio access network (RAN) to improve system performance w.r.t. different types of services envisioned in 5G, i.e. enhanced Mobile Broadband (eMBB), mMTC and URLLC. It is worth noticing that the term "sidelink (SL)" in the 3rd Generation Partnership Project (3GPP) refers to a direct radio link between two arbitrary devices over an interface called PC5 and, therefore, currently comprises the low radio protocol layers, i.e. up to the packet data convergence protocol (PDCP) layer. However, the term "D2D communication" normally refers to an end-to-end (E2E) application. In addition, another focus of this thesis is the exploitation of the relevant context information, e.g. channel state information (CSI), user equipment (UE) location, UE-specific QoS requirement, service priority, battery level, [SKMS12, MKSS11, MSKS11, KRSS14, MSC+13] that is beneficial for enhancing the performance of D2D operation.

In the eMBB use case of 5G, the mobile data volume per area is expected to be 1000 times and the user data rate 10 to 100 times higher than those provided by legacy 4G system. In order to fulfill these requirements, different techniques have been proposed for 5G. For instance, the feasibility of dedicating more spectrum resource to operators for providing eMBB services has been discussed, especially in the frequency range up to 60 GHz [LTRa+18], which is part of the millimeter wave band. Moreover, network infrastructure densification has also been considered to increase the spatial reuse of system resources and, therefore, improve the data volume per area and the user data rate [CQH+16, GSA16]. However, these solutions require more financial investment from mobile network operators (MNOs). In comparison, D2D communication with an underlaying mode has the potential to enable a spatial reuse of spectral resources without purchasing new spectrum or access points. In the D2D underlaying mode, D2D communication operates on the same resource as cellular communication at the same time [MRA17]. Specifically, it is favored to reuse the cellular uplink resource compared with the cellular downlink resource, since the uplink resource is often less utilized and the cellular pilot and synchronization signals are always transmitted in the downlink [LAGR14]. On the other hand, a critical design problem of D2D communication in the underlaying mode is the mutual interference between the two links using the same resource. In this thesis, a network-

controlled sidelink communication taking account of context information is developed to manage the radio resource allocation and mitigate the mutual interference. In 5G, context information needs to be identified from different sources, e.g. a UE and a BS, and it will act as a key to support an efficient radio resource management (RRM) [MET17c]. As mentioned before, the exploited context information in this thesis includes CSI, UE location, UE-specific QoS requirement, service priority, battery level, etc.
Currently, there are multiple factors that demand an increased number of connected MTDs: The smart-grid, large-scale environment and structure monitoring, asset and health monitoring, etc [BPW+18]. Since the MTDs in these applications only sporadically transmit small data packets to the network, they do not consume large data volume capacity. However, due to the characteristics of mMTC applications, the QoS requirements are divergent from the human-type services. For instance, it is predicted to have tens of billions of MTDs beyond 2020. Thus, how to provide such a massive connectivity by the cellular network is an important research question. Moreover, the new levels of availability and power efficiency requirements for mMTC pose technical challenges on the RAN. Thus, 3GPP has conducted research to evolve the legacy 4G system to cover this new type of services. For example, in order to reduce the implementation cost of an MTD, 3GPP has proposed a new type of UE, i.e. the UE category 0 in [3GP16d], where both bandwidth and peak data rate are reduced. Moreover, the costs of the power amplifier (PA) can be reduced by limiting the maximal transmit power [3GP13c]. However, any reduction in transmit power can introduce a negative impact on the network coverage. In order to maintain a good coverage, other technologies, e.g. narrow band transmission and massive transmission time interval (TTI) bundling [3GP15a], have been proposed, but they introduce a large battery drain at an MTD. Therefore, in this thesis the author investigates how to improve network coverage and meanwhile improve power efficiency for mMTC by exploiting sidelink communication.
As another significant aspect envisioned in 5G, Vehicle-to-Everything (V2X) communication is considered as a typical application of URLLC due to its required low latency and ultra-high reliability. Since V2X communication also refers to a local information exchange procedure in most of the cases, it can be deemed as a special type of D2D commu-

nication. However, the relevant users in V2X communication are only traffic participants, and its main goal is to improve traffic safety and efficiency beyond 2020 [MET13a]. To be more specific, V2X communication includes different communication profiles such as Vehicle-to-Vehicle (V2V), Vehicle-to-Pedestrian (V2P), Vehicle-to-Infrastructure (V2I) and Vehicle-to-Network (V2N) communications. Since the legacy 4G system is not capable of providing a reliable V2X communication, e.g. reliability of 99.999%, with a low packet E2E latency, e.g. 5 ms, both academia and industry are working together to design the 5G system to support different automated driving applications including vehicle platooning, advanced driving, remote driving, and sensor information sharing [SJD+18]. For instance, efforts have been made in recent years to offer V2X communication by the 802.11p protocol standardized by the Institute of Electrical and Electronics Engineers (IEEE) [IEE10]. Since the IEEE 802.11p protocol operates in a decentralized mode, there is no central entity to coordinate the transmissions of different users. In order to support this decentralized mode, the medium access control (MAC) layer of the IEEE 802.11p protocol applies a carrier-sense multiple access with collision avoidance (CSMA/CA) scheme that cannot guarantee strict reliability in the case of a high system load. Comparing to the decentralized mode, the 5G system can coordinate the transmissions from different users and, therefore, it has the potential to offer better reliability. So far, two air interfaces have been considered as candidates to support V2X communication in the cellular network, i.e. the LTE-Uu and the PC5 interfaces. In this thesis, the feasibility of applying these two interfaces in V2X communication is examined. This work also improves the communication reliability by proposing a network-controlled two-hop V2X communication and a multi-radio access technologies (multi-RAT) concept.

It is worth noticing that 3GPP has standardized four transmission modes (mode 1-4) for the PC5 interface to support different types of proximity-based services (ProSe). For instance, to support the public safety services, the structure of the physical uplink shared channel (PUSCH) is re-used for sidelink transmission modes 1 and 2. However, for the vehicular applications to improve traffic safety and efficiency, both modes 1 and 2 are not suitable [MMG17] due to the high user mobility. Thus, to cope with V2X communication, the V2X sidelink transmission modes 3 and 4 have been standardized in 3GPP by modifying the relevant functions in modes 1 and 2. For instance, additional demodulation reference signals (DMRS) have been added to the physical layer to handle the high Doppler associated with the relative speeds of up to 500 km/h [3GP18d]. Specifically, both transmission modes 1 and 3 require a radio resource being scheduled by the network, while modes 2 and 4 have been designed to facilitate a UE-autonomous radio resource selection approach.

1.2 Objectives of the thesis

In practical terms, the following list depicts the prime aspects that this thesis copes with:

1. **Network-controlled sidelink communication in the eMBB use case to offload network traffic**: To develop efficient RRM algorithms for allocating radio resources to both the cellular and underlaying D2D links. In order to optimize system performance, context information should be collected with a reasonable signaling effort and correspondingly taken into account by the RRM algorithms.

2. **Applying network-controlled sidelink communication in the mMTC use case to improve network coverage and device power efficiency**: To simultaneously enhance UE power efficiency and extend network coverage by employing sidelink communication among MTDs that are in the proximity of each other. The principal issues to be solved here include how to assign different transmission modes (TMs), e.g. cellular, relay and sidelink modes, to different MTDs and how to design the corresponding control-plane (CP) procedures such as signaling schemes.

3. **Network-controlled V2X communication to provide a low E2E latency and high reliability**: To inspect V2X communication under the scope of cellular technologies. In a cellular network, V2X communication can be realized by either transmitting data packets through network infrastructures, i.e. via the LTE-Uu interface, or a direct V2X communication, i.e. via the PC5 interface.

Thus, both approaches need to be investigated to analyze their pros and cons. Moreover, since the ultra-high reliability requirement in V2X communication cannot always be satisfied by using a single-RAT or a single-hop V2X communication, this work should enhance the communication reliability with new technical proposals, e.g. a multi-RAT V2X communication and a two-hop V2X communication.

Last but not least, in order to obtain a quantitative understanding of the achievable performance gains, the proposed schemes should be implemented and evaluated in a simulation platform which is able to reflect the real world scenarios.

1.3 Contributions of the thesis

The contributions of D2D communication in this work can be structured w.r.t. different types of services in 5G, i.e. eMBB, mMTC and V2X. Please note that the direct communication mode is first introduced in the 3GPP release 12, and it is also considered as a technology for 5G network [MET14a]. Thus, the developed technology in this thesis provides a technical design of D2D communication for the 5G RAN. The concepts and algorithms described in this thesis represent the main contributions of the author to the European FP7 project METIS, Horizon 2020 project METIS-II, and the German BMBF project 5G NetMobil developed between 2013 and 2018. The majority of this research work was conducted, when the initial 5G standard had not been defined. In order to inspect the D2D communication from a system level perspective, the author cannot develop every function and, therefore, some concepts from the legacy cellular LTE network are reused by this work as reference in evaluation. For instance, the link-level performance of the LTE air interface is reused in this thesis. For investigations on the performance improvement of D2D communication, the Long Term Evolution (LTE) technology, e.g. 3GPP Release 12, is considered as a reference. In essence, the research activities carried out in this thesis focus on the 5G CP design, and the LTE functionalities have been reused to generate the user-plane (UP) performance. Moreover, since 5G targets at a control and user plane split that allows an independent evolution of the CP and UP functions [MET17b], the developed concepts and algorithms in the

CP can also be applied in future cellular systems with further evolution on UP functions.
Moreover, since the evaluation methodologies for the legacy 4G system were proposed to evaluate the performance of human-type services, the evaluation framework is adjusted and extended to cover the new service types in this thesis, e.g. mMTC and V2X. In the following, the detailed contributions of this thesis will be categorized w.r.t. the service types.

1.3.1 Network-controlled sidelink communication to offload network traffic

This work

- proposes mathematical models to optimize system performances w.r.t. the number of established D2D links, system capacity, and user satisfaction,
- develops heuristic RRM algorithms with low complexity to solve the constructed optimization problems,
- improves the efficiency of the RRM algorithms by exploiting context information, e.g. channel state information, UE-specific data rate requirement, service priority, and user location,
- designs corresponding signaling schemes to support the proposed sidelink communication and collect the required context information with a reasonable signaling effort.

1.3.2 Applying network-controlled sidelink communication in the mMTC use case

This work

- applies multiple clustering schemes for grouping mMTC devices to ensure any two mMTC devices forming a D2D link are not far away from each other,
- develops a smart TM selection approach to configure relay and remote UEs in each cluster by using context information, e.g. sensor location and the real-time battery level,

- designs signaling schemes to both collect the relevant context information and enable the proposed sidelink communication that improves the network coverage and energy consumption of MTDs.

1.3.3 Network-controlled V2X communication

This work

- introduces how to apply a cellular V2X communication through network infrastructures, i.e. via the LTE-Uu interface, and a direct V2X communication, i.e. via the PC5 interface,
- develops the methodology to evaluate the V2X communication,
- evaluates the performance of both the cellular V2X communication via the LTE-Uu and the direct V2X communication via the PC5 (the performance can be considered as a baseline to evaluate new technical proposals in the future),
- proposes a network-controlled two-hop V2X communication via the PC5 interface to increase the communication reliability in the case that a large V2X communication range is required,
- optimizes the resource allocations to the first hop and the second hop in order to achieve a maximal packet transmission range for the proposed two-hop V2X communication,
- explains how to collect and exploit context information to make the proposed two-hop V2X communication more efficient,
- designs a multi-RAT concept to enhance V2X communication by transmitting data packets via both the LTE-Uu and the PC5 interfaces,
- proposes to equip base stations (BSs) with the capability to receive packets from UEs via the PC5 interface in order to achieve a better robustness in uplink transmission.

1.3.4 Performance evaluation of D2D and V2X communications

This work

- proposes an evaluation framework for ProSe w.r.t. different service types, i.e. eMBB, mMTC and V2X,
- implements the system-level simulators in MATLAB to derive the performances of the proposed technologies,
- quantitatively evaluates the performance gains of D2D and V2X communications in terms of different key performance indicators (KPIs), e.g. system capacity, number of established D2D links, device battery life, network coverage, packet E2E latency, etc,
- validates the benefits of applying direct D2D operation, including an increased spectral efficiency, an enlarged network coverage, a better UE power efficiency, a lower E2E latency, and higher communication reliability.

1.4 Relevant publications

The contributions of the thesis were published in various peer-reviewed magazine, journal and conference papers, and also in two book chapters. The publication list for the topic on exploiting a network-controlled sidelink communication to offload network traffic includes [JKK$^+$14, JKKS14, JWKS17, SJD$^+$18]. The work related with the network-controlled sidelink communication in the mMTC use case is published in [JLS17, JHLS17, SJD$^+$18]. In addition, [JLWS17, JWHS17, JWHS18b, JDWS18, JWHS18a] contain the relevant work of the network-controlled V2X communication. Moreover, part of the 5G evaluation framework w.r.t. different use cases has been captured in [MFMSJ16]. Last but not least, the publications of the author in [SKJ$^+$15, KKJ$^+$15, JT13, TJ14, WKJS17, RJKS15, SJMS13, KWJS18, HJS18] also comprehensively contribute to a better understanding of the topics addressed in this thesis.

1.5 Organization of the thesis

This thesis is organized as follows:

1. The author first presents the state-of-the-art evaluation methodology of the legacy 4G system in Chapter 2 and discusses how to extend and modify the legacy evaluation framework for assessing

the technical D2D and V2X communication proposals in the 5G system. After that, the author provides detailed descriptions regarding the simulation models that will be used and implemented in the following chapters to evaluate the technical proposals.

2. In Chapter 3, a context-aware network-controlled sidelink communication concept is introduced to offload traffic from cellular communication to local communication. In this chapter, the author designs multiple RRM algorithms to optimize network performance w.r.t. different targets, e.g. the maximal number of established D2D links and system capacity. Moreover, in order to offer the eMBB services with different QoS requirements, e.g. data rate and priority level, to the mobile users, context information is taken into consideration by the proposed RRM algorithms. Please note that at the beginning of this chapter only CSI is used to derive the context-aware RRM algorithm. Though a context-aware scheme usually does not only apply the CSI, it is exploited as a starting point, and other context information, e.g. UE location, UE data rate requirement, and service priority, is taken into account by the RRM algorithms in the later part of this chapter. In addition, to support the RRM algorithms and collect the required context information, signaling schemes in both single-cell and multi-cell scenarios are designed in this chapter. Further, the performance of the proposed algorithms are simulated, and their results are compared with other baseline schemes.

3. Chapter 4 deals with how to exploit sidelink communication to improve the UE energy consumption and the network coverage simultaneously in the mMTC use case. In order to assist a BS for pairing a remote sensor with an appropriate relay sensor, the sidelink configuration task is decomposed into two sub-tasks, i.e. device clustering and TM selection in this chapter. Correspondingly, three essential signaling schemes are also introduced to initialize sensor attachment, update TM, and transmit uplink report via sidelink communication. Finally, simulation results obtained from both the urban and the rural areas are given and analyzed.

4. At the beginning of Chapter 5, the two different approaches of realizing V2X communication using the LTE-Uu air interface and the

PC5 air interface are investigated. Following that, a concept of network-controlled two-hop V2X communication via the PC5 interface is proposed to meet the high-reliability requirement of the V2X communication in a large communication range. In this part, the author also shows that the transmission range of data packets can be maximized by adjusting the amount of resources allocated to the two different hops. Moreover, context information is collected and applied to configure relay vehicles for the second hop transmission. In addition, a multi-RAT scheme is also developed in this chapter to achieve a diversity gain, since using one single-RAT might not always fulfill the reliability requirement. This proposal enables data transmission through both the cellular network infrastructures, i.e. via the LTE-Uu, and a direct V2X communication, i.e. via the PC5 interface. In the last part of this chapter, the numerical results provide a deep insight into the performance comparison of the different V2X communication technologies.

5. In Chapter 6, the author concludes the contributions of this thesis and outlines the possible areas of future work on D2D and V2X communications.

Chapter 2
Evaluation Framework and Methodology

2.1 Evaluation methodology for 4G

The legacy 4G system was designed to mainly serve human-driven traffic, and it provides a better user experience compared with the previous cellular systems, e.g. 2G and 3G. The performance metrics typically defined for a 4G system include cell spectral efficiency, peak data rate, bandwidth scalability, cell-edge user spectral efficiency, latency, handover interruption time, and VoIP capacity [IR08b].
In order to set a common criterion for different organizations to evaluate the performance of different 4G technical candidates, the ITU Radiocommunication Sector (ITU-R) has come up with a guideline document in [IR08a] with considerations from different perspectives, e.g. vendors, network operators, device manufacturers, and service and content providers. Moreover, as the most interesting part for simulation experts, [IR08a] also describes the system simulation procedures, e.g. deployment model, traffic model, mobility model and channel model, and the evaluation methodologies to derive the performance of different KPIs.
In principle, the KPIs can be categorized into three main classes by looking at the approaches applied for their evaluation [WWLS10, SZJ+09, Ahm10]:

- KPIs to be analytically evaluated, e.g. intra-/inter frequency handover interruption time, peak data rate, peak spectral efficiency, and control/user plane latency,
- KPIs to be inspected by examining the protocol designs and checking if certain features are supported, e.g. inter-system handover,
- KPIs to be evaluated with simulation, e.g. cell/cell-edge user spectral efficiency, VoIP capacity, mobility.

2.1.1 Analysis-based and inspection-based KPI evaluation

The performance of the analysis-based KPIs can be calculated based on the RAT specifications [Oss11]. For instance, the peak data rate can

be calculated by extracting the overhead, e.g. synchronization and reference signals, from the maximal data bits per time unit. Moreover, in order to check whether the CP and UP latency of a technical proposal comply with the requirements specified for International Mobile Telecommunications-Advanced (IMT-Advanced) systems, the calculations for both the frequency-division duplex (FDD) and the time-division duplex (TDD) modes are given in [SKC10]. In addition, to inspect how well a cellular system can support user mobility, the handover interruption time should be derived. This can be done by dividing the whole handover procedure to different technology-specific delay components and computing the summation of their values [SKC+11]. Since a relay concept has been studied and introduced in 3GPP Release 9, the cell spectral efficiency of a relay-enhanced cell deployment scenario can also be calculated using the methodology defined in [BAS10].
In the other cases where an inspection method is applied, the corresponding KPIs are evaluated by inspecting the system design information, e.g. protocol and architecture design, and the assessment of these KPIs requires only a Yes/No answer. For instance, by inspecting the network interoperability design between 4G and 3G, it can be seen whether an inter-system handover is feasible to avoid dropping of a call when a mobile device enters the coverage area of a 3G system from a 4G system.

2.1.2 Simulation-based KPI evaluation

In the simulation-based approach, the KPIs need to be derived by running a simulator taking account of the different features in the relevant protocol layers, e.g. modulation and coding scheme (MCS), hybrid automatic repeat request (HARQ), and scheduling. Moreover, in order to reduce computational complexity, a mapping table from signal-to-interference-plus-noise ratio (SINR) to block error rate (BLER) should be generated from a link-level simulator and integrated into a system level simulator [HS13]. Moreover, to precisely reflect the characteristics of a radio link, e.g. frequency-selective fading, a link-to-system (L2S) level mapping needs to be accurately formulated. In literature [BAS+05], two metrics have been widely used to construct the L2S level mapping, i.e. the exponential effective SINR metric (EESM) and the mutual information effective SINR metric (MIESM). Due to a reduced complexity to

effectively reproduce the link-level performance [HS13], the MIESM has been considered as preferable and applicable to a large class of MIMO-OFDM transmission techniques [Ahm10, BAS+05]. To provide a good insight into the system-level simulator, [MSMOC11] takes account of the relevant functionality, e.g. network layout, channel model, L2S mapping, and characters of a BS and a mobile terminal.
As one of the IMT-Advanced systems, the evaluation methodology of the IEEE 802.16m protocol has been comprehensively reported in [SZJ+09, Jai08] with a systematic introduction on how to model the corresponding technical features. For instance, a sound presentation on L2S mapping by using the effective SINR is provided and, therefore, a prediction of the instantaneous link-level performance can be enabled with a low complexity. Moreover, the detailed evaluation guidelines are also demonstrated to assist the inspection on link adaptation, HARQ, and radio resource scheduling algorithm.
In particular, as the multi-antenna technology is considered as an essential technical component in 4G system, the mathematical models to evaluate the performance, e.g. spectral efficiency, of a multiple-input and multiple-output (MIMO) scheme have been developed in [MKM+10, STB10].
In order to facilitate a fair comparison among the technical proposals from different proponents, ITU-R has also come up with its guideline document in [IR08a] so that different evaluation groups and organizations can independently implement their own platforms. In this document, the evaluation process and simulator setup parameters are provided. Please note that, in order to validate the correctness of the simulation platforms from different developers, a calibration step is necessary, and the basic steps for both the link-level calibration and system-level calibration have been captured and explained in [WWLS10].
In addition, some simulation scenarios have also been proposed by 3GPP in [3GP10]. This document describes the baseline assumptions to evaluate the LTE-Advanced technology, e.g. deployment models, channel models, traffic models, and antenna patterns.
It is worth noticing that an analytical evaluation approach has been proposed in [MBSW11] to calculate the cumulative distribution function (CDF) of SINR values and further map from SINRs to the corresponding data rates. This approach offers an efficient method to verify the results

generated by a system-level simulator and spot out a model imperfection or possible implementation errors.

2.2 Evaluation of beyond 4G technologies and capabilities

The state-of-the-art evaluation methodology of 4G system has its limitations and challenges to evaluate beyond 4G technologies and, therefore, modifications and extensions are required. For example, the channel models to evaluate 4G techniques are only reasonable for certain specific deployment scenarios, e.g. where BSs are deployed above rooftop and UEs are far away from their serving BS. However, these models miss models for the vertical dimension, e.g. elevation angles to both the transmitter (Tx) and receiver (Rx) antennas. Thus, in order to evaluate massive and ultra-dense deployments in 5G, more accurate channel models taking account of the elevation dimension and the corresponding changes in radio propagation are required.
Since wireless systems have evolved over the last years, [LJ12] inspects whether the classical settings of evaluation framework remain still adequate in the light of the rapid advances of wireless communication systems. As the two most commonly used classical settings to model fading dynamics, either the ergodic settings or the quasi-static settings are applied depending on whether the channel coherence time is shorter than the packet duration, i.e. the ergodic settings are applied, or not, i.e. the quasi-static settings are applied. [LJ12] discusses how these settings interact with the features of wireless systems w.r.t. link adaptation, HARQ, wideband signaling and operating point. Furthermore, it implements a simple link-level simulator and provides a good insight for modeling the link-level simulation.
In order to inspect the importance of an accurate radio propagation modeling, [ESE13b] derives the system performances by using three different models for elevation angular spread. The numerical results demonstrate a performance difference up to 25% w.r.t. average user throughput by using different angular spread models. Thus, it can be observed that an appropriate modeling of radio propagation is essential to yield valid simulation results.

In [ESE13c], the approach to incorporate the three-dimensional (3D) aspects in channel modeling is discussed. In particular, it extends the legacy 4G channel models to account for the heights of BSs, UEs, and buildings in a Manhattan grid environment where both the locations of streets and buildings are properly modeled.

In [ESE13a], 3D channel models for outdoor signal propagation are proposed. In this proposal, radio signals mainly travel along two propagation paths, i.e. around buildings and above rooftops. As a basis, the urban micro (UMi) model and the urban macro (UMa) model proposed by 3GPP in [3GP10] are respectively applied for the around building propagation and above rooftop propagation. In addition to that, [ESE13a] also proposes a line-of-sight (LOS)/non-line-of-sight (NLOS) propagation model taking the heights of the Tx, the Rx and buildings into account. Compared with the current ITU-R channel models, where a stochastic procedure is used to determine the LOS/NLOS condition, this proposal is able to remove the discontinuous jumps of pathloss values over the geographical locations due to the switching of LOS/NLOS states. However, additional computational complexity is expected in this approach, since the directions and locations of streets, as well as angular characteristics need to be taken into account by the simulator.

Moreover, [NN13] describes a way to create 3D channel models by using the 2D ITU-R channel models as a basis. In that work, both the median elevation angle and the angle spread at a BS and a UE are detailed.

Since spectrum is deemed as an expensive and critical resource in cellular systems, how to optimize the utilization of the precious spectral resource has attracted great research interest. In the last years, the concept of densification, deploying low power nodes or femto cells has been expected to dramatically increase spectral efficiency. Thus, in order to provide an evaluation baseline for femto cells, [3GP10, HHCC13] describe the corresponding link-level simulation parameters, network deployment, and performance metrics.

Moreover, due to the deployment of small cells, a world with more BSs than UEs can be foreseen for future mobile systems [Mal12]. With a heterogeneous network deployment, the small cells are often lightly loaded, while the macro cells are heavily loaded. Therefore, the traffic congestion problem plays an important role in determining the achieved system capacity. Therefore, in [And13], it is recommended to stop indicating system performance with SINR distribution or spectral efficiency as for the 4G system. Instead, a rate distribution or area spectral efficiency should be exploited to better capture the real system performance in a heterogeneous network. Moreover, it is also proposed to model the deployment of femto cells by a Poisson point process in order to well describe the statistics.

In order to provide a consistent definition, specification, and evaluation of the IMT for 2020 and beyond (IMT-2020) radio technologies, [IR15] describes the framework of the future system development associated with the 5G use cases, and defines the minimum technical performance requirements. In this report, eight key capabilities of the IMT-2020 system have been compared with those of IMT-Advanced to show the capability enhancement of 5G, e.g. peak data rate, user experienced data rate, spectrum efficiency, mobility, latency, connection density, network energy efficiency, and area traffic capacity. The defined technical performance requirements target at fulfilling the IMT-2020 objectives. As the main standardization body for 5G, 3GPP has also performed a study on 5G service requirements. In [3GP16b], 74 use cases have been identified and grouped w.r.t. their service requirements. In addition, [3GP16c] has provided a detailed study on the 5G deployment scenarios and performance requirements. In this report, the methodology to evaluate the relevant KPIs has been described.

2.3 Simulation guidelines for this thesis

In order to illustrate the models used in this work for evaluating the proposed D2D technologies, detailed information is provided in this section. As mentioned before, since D2D is applied in different use cases which are clustered w.r.t. the service requirements and application scenarios, i.e. eMBB, mMTC, and URLLC, the models and parameters to evaluate a specific D2D communication technology can be varied in different use cases.

2.3.1 KPIs and their evaluation

The following KPIs [MET16, MET13b] are used in this thesis to evaluate the effectiveness of the technical proposals.

Cell throughput

The cell throughput is the aggregation of the throughput of all users in a cell. Thus, the cell throughput can be computed as the total amount of successfully decoded information bits divided by the time.

Network coverage

The network coverage is the geographical area where an access point and a UE can communicate with a predefined minimum QoS. Please note that the downlink and uplink communications in a cellular network can have different coverage due to the different maximal transmit powers at a BS and a UE.

Battery life of mMTC devices

The battery life of an mMTC device is the operating time with a predefined battery capacity only. Please note that this KPI should be derived without any battery replacement. In this work, the power leakage effect is not taken into account.

CP latency

CP latency should be calculated as the transition time for a UE to enter the RRC_CONNECTED state from the RRC_IDLE state [IR08b].

UP end-to-end latency

The UP E2E latency, or one trip time (OTT) latency, refers to the time taken for a data packet to be successfully transmitted from a source to its destination. The measurement reference point, in this thesis, is the interface between Layers 2 and 3.

Packet reception ratio

Packet reception ratio (PRR) is a critical KPI in V2X communication since it is related to the communication reliability. It should be calculated as a ratio of X to Y, where X is the number of packets successfully received within a certain time interval, e.g. 100 ms in this work, and Y represents the number of packets which should be received.

2.3.2 Deployment scenarios

As mentioned before, new use cases should be taken into account by the evaluation framework in 5G. In the following, the details on both synthetic and realistic deployment scenarios used in this thesis are presented.

2.3.2.1 Synthetic deployment scenario

In a synthetic deployment scenario, no specific topographical details will be taken into consideration. For instance, no buildings and roads will be placed in the environment model, and UEs are randomly and uniformly distributed on a two-dimensional (2D) plane in the coverage area. In this case, a regular hexagonal cell layout is typically assumed for each cell and BS locates at the center of the hexagons. Thanks to this simplified environment model, the complexity for implementing and setting up a simulator can be efficiently reduced. However, modeling accuracy will also be lost due to that. At first, since UEs are not distinguished regarding whether they locate indoor or outdoor, traffic and mobility profiles fitting a UE-specific scenario cannot be correspondingly attached to each UE. Thus, velocity and data service types are randomly selected to describe a UE's movement and its requested data service in the simulator. Secondly, the channel model in this case cannot take account of the influence of a real environment, e.g. signal reflection and refraction. This results in a random selection of LOS or NLOS propagation scenarios for each geographical location and, therefore, a big discontinuous jump regarding pathloss values between two geographically neighboring locations can be experienced.

2.3.2.2 Realistic deployment scenario

As mentioned in Section 2.2, it is critical to introduce a realistic scenario in 5G for evaluation purposes. The 3GPP technical document [ESE13b] demonstrates the need to precisely reflect the real world in a valid simulator in order to obtain reasonable observations on system performance. Thus, to capture the dynamic traffic variations in a dense urban environment w.r.t. both space and time, the Madrid grid model proposed by the METIS project [MET13b] is implemented in this work as a realistic scenario.

As seen in Figure 2.1, the Madrid grid model can be used to describe an urban environment with 3D visualization. It covers an area of 387

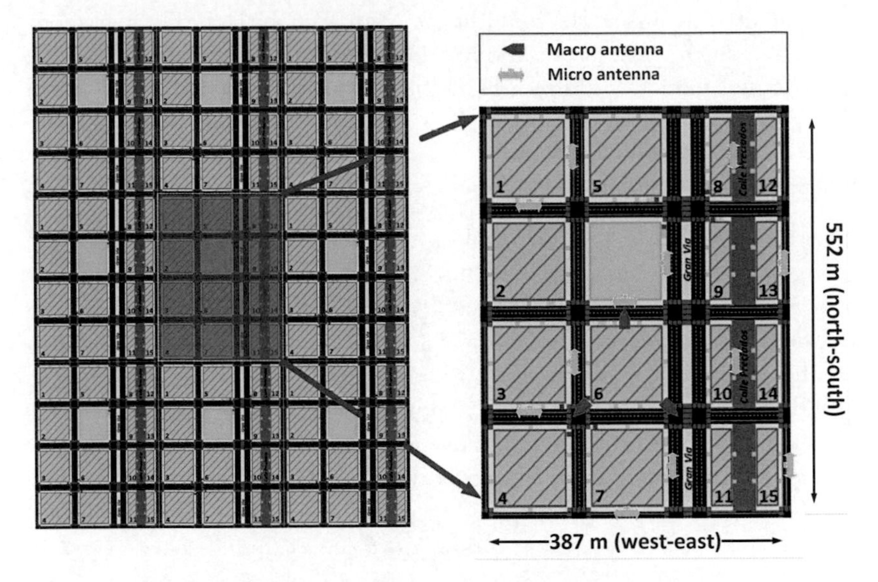

Figure 2.1: Environment and deployment models in Madrid grids

m (east-west) and 552 m (south-north) for each grid. One central park and 15 buildings with different heights and sizes are contained in each Madrid grid. Moreover, the heights of different buildings are distributed randomly and uniformly between 8 and 15 floors with a height of 3.5 m per floor. A single macro station is deployed on the roof of building 6 and operates in three sectors. Directional antennas with 120-degree difference from each other in the horizontal plane are used. In addition to the macro BS, micro BSs can also be deployed in this model to set up a heterogeneous network. It is assumed that each micro BS has two sectors and the antenna of each sector points towards one side of the main street. To achieve a specific cell radius and avoid cell border effects, multiple replicas of Madrid grids can be wrapped around in a system-level simulator.

2.3.3 Channel models

In order to capture the radio propagation characteristics, the channel models for both the synthetic and realistic scenarios are considered and implemented in this work. In addition, compared with the channel models in 4G which consider a radio link between an access point and a UE, the channel model of a direct D2D link between two UEs is also taken into account in this thesis.

2.3.3.1 Pathloss models for synthetic scenarios

2.3.3.1.1 Cellular link The pathloss models proposed in [IR08a, 3GP10] are used to characterize cellular links in the synthetic scenario and they are given in Table 2.1. Later, the LOS and NLOS models in urban macro (UMa) scenario are used in Chapter 3 to model the radio channel in the synthetic scenario. While the LOS and NLOS models in rural macro (RMa) scenario are applied in Chapter 4. Please note that the carrier frequency f_c in this table should have a unit of GHz and the signal propagation distance d is in m. Moreover, the distance of the breaking point is calculated as

$$d_{\mathrm{BP}} = 4h'_{\mathrm{BS}}h'_{\mathrm{UT}}f_c/c, \tag{2.1}$$

where h'_{BS} and h'_{UT} are the effective antenna heights of the BS and the UE, respectively, and c is the speed of light. The relationship between the actual antenna heights, e.g. h_{BS} and h_{UT}, and the effective antenna heights can be computed as follows:

$$h'_{\mathrm{BS}} = h_{\mathrm{BS}} - 1.0 \text{ m}, \qquad h'_{\mathrm{UT}} = h_{\mathrm{UT}} - 1.0 \text{ m}. \tag{2.2}$$

In addition, the probability of an LOS propagation from a Tx to its Rx can be computed as shown in Table 2.3.

2.3.3.1.2 D2D link The outdoor D2D pathloss models shown in [IR08a, 3GP13b] are applied for the D2D links in urban and rural areas. Later, the LOS and NLOS models in UMa are used in Chapter 3 to model the direct radio channel in the synthetic scenario. While the LOS and NLOS models in RMa are applied in Chapter 4 for modelling the direct link between two sensors in the rural area. In addition to that, the Okumura-Hata model is used to reflect the propagation characteristics for the direct V2X communication in a highway scenario in Chapter 5. These pathloss models are described in Table 2.2, where the carrier frequency f_c is given in GHz and the distance d in m. Last but not least, the LOS probability for a D2D link is given in Table 2.3.

Table 2.1: Pathloss models for cellular links [IR08a]; f_c in GHz; all distances in m

Scenario	Pathloss (PL)/dB	Shadow fading std/dB	Applicability range, antenna height
UMa LOS	$PL = 22.0 \log_{10}(d) + 28.0 + 20.0 \log_{10}(f_c)$	$\sigma = 4$	10m$< d < d_{BP}$
	$PL = 40.0 \log_{10}(d) + 7.8 - 18.0 \log_{10}(h'_{BS})$ $-18.0 \log_{10}(h'_{UT}) + 2.0 \log_{10}(f_c)$	$\sigma = 4$	$d_{BP} < d <$ 10000m h_{BS} =25m, h_{UT} =1.5m
UMa NLOS	$PL = 161.04 - 7.1 \log_{10}(W) + 7.5 \log_{10}(h)$ $-(24.37 - 3.7\ (h/h_{BS})^2) \log_{10}(h_{BS}) + (43.42$ $-3.1 \log_{10}(h_{BS}))(\log_{10}(d) - 3) + 20 \log_{10}(f_c)$ $-(3.2\ (\log_{10}(11.75\ h_{UT}))^2 - 4.97)$	$\sigma = 6$	10m$< d <$ 5000m h =avg. building height W =street width h_{BS} =25m, h_{UT} =1.5m, W =20m, h =20m 5m$< h <$50m, 5m$< W <$50m 10m$< h_{BS} <$150m, 1m$< h_{UT} <$10m
RMa LOS	$PL_1 = 20 \log_{10}(40\pi\ d\ f_c/3)$-min$(0.44\ h^{1.72}, 14.77)$ $+0.002 \log_{10}(h)\ d$+min$(0.03\ h^{1.72}, 10) \log_{10}(d)$	$\sigma = 4$	10m$< d < d_{BP}$
	$PL_2 = PL_1(d_{BP}) + 40 \log_{10}(d/d_{BP})$	$\sigma = 6$	$d_{BP} < d <$10000m h_{BS} =35m, h_{UT} =1.5m, W =20m, h =5m (Applicability ranges of h, W, h_{BS}, h_{UT} are the same as in UMa NLOS)
RMa NLOS	$PL = 161.04 - 7.1 \log_{10}(W) + 7.5 \log_{10}(h)$ $-(24.37 - 3.7\ (h/h_{BS})^2)\log_{10}(h_{BS}) + (43.42$ $-3.1 \log_{10}(h_{BS}))(\log_{10}(d) - 3) + 20 \log_{10}(f_c)$ $-(3.2\ (\log_{10}(11.75\ h_{UT}))^2 - 4.97)$	$\sigma = 8$	10m$< d <$ 5000m, h_{BS} =35m, h_{UT} =1.5m W =20m, h =5m (The applicability ranges of h, W, h_{BS}, h_{UT} are the same as for UMa NLOS)

Table 2.2: Pathloss models for D2D links in synthetic scenarios; d in m [IR08a, 3GP13b]

Scenario	Pathloss/dB	Shadow fading std/dB
UMa and RMa LOS	$PL = 92.45 + 20\log_{10}(f_c) + 20\log_{10}(d/1000)$,	$\sigma = 7$, log-Rayleigh distribution
UMa and RMa NLOS	$PL = 144.5 + 45\log_{10}(f_c) + 40\log_{10}(d/1000)$ $+L_{\text{urban}}$, $L_{\text{urban}} = 0,\ 2.3$ dB for rural, and dense urban areas respectively	$\sigma = 7$, log-normal distribution
Highway	$PL = 126.61 + 15.72\log_{10}(f_c) - 4.78\,(\log_{10}(f_c))^2$ $+43.75\log_{10}(d)$	

Table 2.3: LOS probability for D2D links; d in m [IR08a, 3GP13b]

Scenario	LoS probability
UMa cellular link	$P_{\text{LOS}} = \min(18/d, 1) \cdot (1 - \exp(-d/63)) + \exp(-d/63)$
RMa cellular and D2D link	$P_{\text{LOS}} = \begin{matrix} 1, d \leq 10 \\ \exp(-\frac{d-10}{1000}), d > 10 \end{matrix}$
UMa D2D link	$P_{\text{LOS}} = \min(18/d, 1) \cdot (1 - \exp(-d/36)) + \exp(-d/36)$

2.3.3.2 3D channel models for realistic scenarios

As mentioned in Section 2.2, a proper model for channel characterization plays a critical role in performance simulation. Therefore, a ray-tracing-based 3D channel model to represent the large-scale effect is motivated in [MET13b]. With this approach, the radio conditions can be tightly aligned with the environment model. For instance, the LOS/NLOS propagation condition should be selected by checking if a building blocks the straight line between two communication ends. For that, the exact location and dimension of each building need to be reflected in the simulator. Recalling that the LOS/NLOS propagation in synthetic scenarios is mathematically described as a stochastic process in Table 2.3, the proposed 3D channel model requires a higher implementation complexity to improve the modeling accuracy. Thus, to achieve a good compromise between the complexity and the accuracy, the single ray-tracing channel model proposed in [MET13b, OMM16] is used in Sections 3.4, 5.2 and 5.5 for modeling the Macro outdoor-to-outdoor (O2O) LOS/NLOS, Micro O2O LOS/NLOS, and V2X LOS propagation situations. In addition, the Macro outdoor-to-indoor (O2I) [3GP15a] and D2D indoor-to-indoor (I2I) [3GP13a] propagation models proposed by 3GPP are exploited in Section 4.5 to evaluate the technical proposals regarding a massive deployment of sensors. Last but not least, the NLOS propagation model in [MKH11] is implemented for the direct V2X communication in the Madrid grid environment. This model will be used in Section 5.5.

2.3.3.2.1 Macro outdoor-to-outdoor propagation model [MFMSJ16] In this scenario, a macro BS locates on the roof of a building in the Madrid grid environment as shown in Figure 2.1. The pathloss in dB between the BS and an outdoor UE can be calculated as [MET13b]

$$PL = \begin{cases} PL_{\text{fs}} + PL_{\text{rts}} + PL_{\text{msd}}, & \text{if } PL_{\text{rts}} + PL_{\text{msd}} > 0; \\ PL_{\text{fs}}, & \text{if } PL_{\text{rts}} + PL_{\text{msd}} \leq 0. \end{cases} \tag{2.3}$$

Equation (2.3) shows that the pathloss value is the summation of three components, i.e. the free space loss PL_{fs}, the diffraction loss from rooftop to the street PL_{rts}, and the multiple screen diffraction loss PL_{msd}. In this section, these three components are in dB. In the following, the calculation of the different components will be discussed. Please note

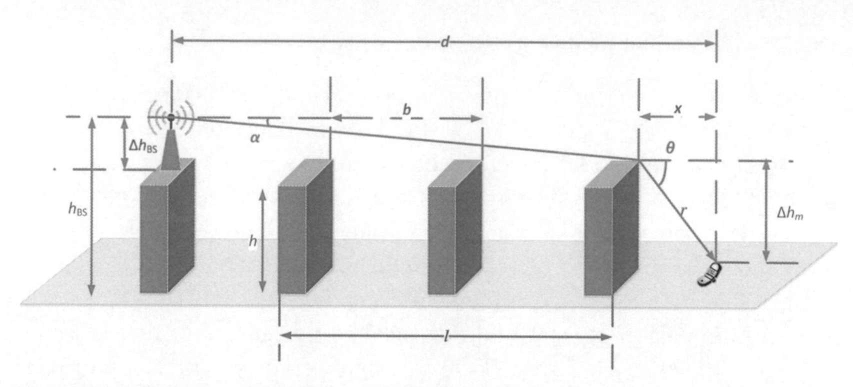

Figure 2.2: Channel model for macro O2O [MFMSJ16] (h_{BS}, Δh_{BS}, h, b, d, l, x, r, and Δh_{m} are in m; while α and θ are in radian)

that the different geographical terms and their units, e.g. h_{BS}, Δh_{BS}, α, h, b, d, l, x, θ, r, and Δh_{m}, are illustrated in Figure 2.2.
The free space loss PL_{fs} can be calculated as

$$PL_{\mathrm{fs}} = -10\log_{10}(\frac{\lambda}{4\pi d})^2, \tag{2.4}$$

where λ is the wavelength. The diffraction loss PL_{rts} due to the diffraction from a rooftop to a street is computed as

$$PL_{\mathrm{rts}} = -20\log_{10}\left\{\frac{1}{2} - \frac{1}{\pi}\arctan\left[\mathrm{sign}(\theta)\sqrt{\frac{\pi^3}{4\lambda}r(1-\cos(\theta))}\right]\right\}, \tag{2.5}$$

where

$$\theta = \tan^{-1}(\frac{|\Delta h_{\mathrm{m}}|}{x}), \tag{2.6}$$

and

$$r = \sqrt{(\Delta h_{\mathrm{m}})^2 + x^2}. \tag{2.7}$$

In order to calculate the multiple screen diffraction loss PL_{msd} coming from signal propagation over multiple buildings, the length of the path l covered by buildings needs to be compared with the threshold value of d_{s} in m that is calculated as

$$d_{\mathrm{s}} = \frac{\lambda d^2}{\Delta h_{\mathrm{BS}}{}^2}. \tag{2.8}$$

If $l > d_\mathrm{s}$, then

$$PL_\mathrm{msd} = L_\mathrm{bsh} + k_\mathrm{a} + k_\mathrm{d}\log_{10}(d/1000) + k_\mathrm{f}\log_{10}(f_\mathrm{c}) - 9\log_{10}(b), \quad (2.9)$$

where the frequency f_c has the unit MHz and $k_\mathrm{f} = 15(f_\mathrm{c}/925 - 1)$ is proposed for metropolitan centers, and the loss term L_bsh in dB depends on the base station height

$$L_\mathrm{bsh} = \begin{cases} -18\log_{10}(1 + \Delta h_\mathrm{BS}), & \text{if} \quad h_\mathrm{BS} > h; \\ 0, & \text{if} \quad h_\mathrm{BS} \leq h. \end{cases} \quad (2.10)$$

Moreover, in Equation (2.9), k_a and k_d are calculated as

$$k_\mathrm{a} = \begin{cases} 54, & \text{if} \quad h_\mathrm{BS} > h; \\ 54 - 0.8\Delta h_\mathrm{BS}, & \text{if} \quad h_\mathrm{BS} \leq h \quad \text{and} \quad d \geq 500; \\ 54 - 1.6\Delta h_\mathrm{BS} d/1000, & \text{if} \quad h_\mathrm{BS} \leq h \quad \text{and} \quad d < 500, \end{cases} \quad (2.11)$$

and

$$k_\mathrm{d} = \begin{cases} 18, & \text{if} \quad h_\mathrm{BS} > h; \\ 18 - 15\frac{\Delta h_\mathrm{BS}}{h} & \text{if} \quad h_\mathrm{BS} \leq h, \end{cases} \quad (2.12)$$

respectively.
On the other hand, if $l \leq d_\mathrm{s}$, the PL_msd is calculated as

$$PL_\mathrm{msd} = -10\log_{10}({Q_\mathrm{m}}^2), \quad (2.13)$$

where

$$Q_\mathrm{m} = \begin{cases} 2.35(\frac{\Delta h_\mathrm{BS}}{d}\sqrt{\frac{b}{\lambda}})^{0.9}, & \text{if} \quad h_\mathrm{BS} > h; \\ \frac{b}{d}, & \text{if} \quad h_\mathrm{BS} \approx h; \\ \frac{b}{2\pi d}\sqrt{\frac{\lambda}{\rho}}(\frac{1}{\vartheta} - \frac{1}{2\pi + \vartheta}), & \text{if} \quad h_\mathrm{BS} < h, \end{cases} \quad (2.14)$$

$$\vartheta = \tan^{-1}(\frac{\Delta h_\mathrm{BS}}{b}), \quad (2.15)$$

and

$$\rho = \sqrt{\Delta h_\mathrm{BS}^2 + b^2}. \quad (2.16)$$

2.3.3.2.2 Micro outdoor-to-outdoor propagation model When a micro BS locates much below the average height of its surround buildings, the signal propagates by reflections between buildings, and the radio propa-

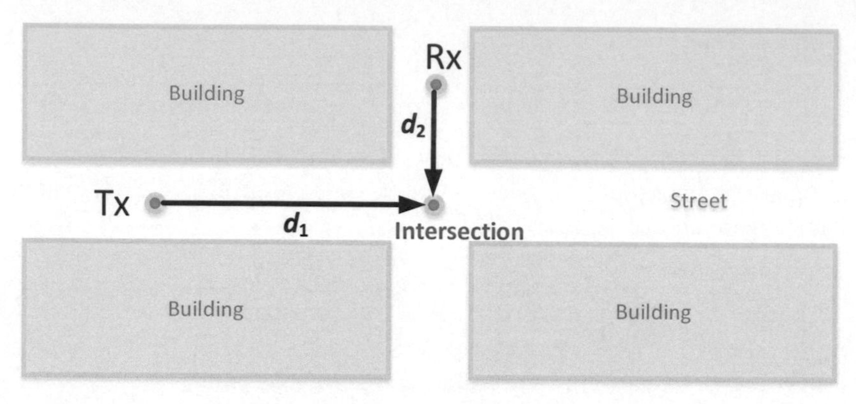

Figure 2.3: Micro O2O NLOS channel model

gation in this case is usually referred as the UMi model [3GP10]. In this case, the simulator needs to first check whether an LOS/NLOS propagation exists between the Tx and Rx, and then calculates the pathloss correspondingly [MET13b].

In case a LOS propagation exists, then

$$PL_{\mathrm{LOS}}/\mathrm{dB} = 40\log_{10}(d) + 7.8 - 18\log_{10}(h'_{\mathrm{tx}}) - 18\log_{10}(h'_{\mathrm{rx}}) + 2\log_{10}(f_{\mathrm{c}}), \tag{2.17}$$

where d is the distance between the Tx and Rx in m, and f_{c} is the carrier frequency in GHz. Additionally, the h'_{tx} and h'_{rx} are the effective antenna heights of the Tx and Rx, respectively.

In the other case, where the two ends of a communication link are located in the perpendicular streets and a NLOS propagation is experienced, the pathloss is a function of d_1 and d_2, which are shown in Figure 2.3, as

$$PL_{\mathrm{NLOS}}/\mathrm{dB} = \min(PL(d_1, d_2), PL(d_2, d_1)), \tag{2.18}$$

where

$$PL(d_k, d_l)/\mathrm{dB} = PL_{LOS}(d_k) + 17.9 - 12.5n_j + 10n_{\mathrm{j}}\log_{10}(d_l) + 3\log_{10}(f_{\mathrm{c}}), \tag{2.19}$$

and

$$n_{\mathrm{j}} = \max(2.8 - 0.0024d_k, 1.84). \tag{2.20}$$

Finally, please note that if the Tx and its Rx locate in parallel streets and there is no LOS between them, the pathloss is assumed to be infinite.

2.3.3.2.3 Macro outdoor-to-indoor propagation model In an urban scenario, an mMTC device is often deployed inside a building. However, the macro BS normally locates on the roof of a building, and thus the building penetration loss is part of its pathloss and needs to be considered in the channel modeling. In this sense, the pathloss value in dB for a device located inside a building has been proposed by 3GPP in [3GP15a], and it should be computed as

$$PL = PL_{\mathrm{fs}} + PL_{\mathrm{bpl}}, \tag{2.21}$$

where PL_{fs} represents the outdoor free-space pathloss in dB, as calculated in Equation (2.4). The building penetration loss PL_{bpl} in dB in Equation (2.21) can be calculated as

$$\begin{aligned} &PL_{\mathrm{bpl}}/\mathrm{dB} = \\ &\max(W_{\mathrm{i}} \times N_{\mathrm{p}}, 0.6 \times d_{\mathrm{p}}) - 1.5 \times n_{\mathrm{f}} + PL_{\mathrm{ew}}. \end{aligned} \tag{2.22}$$

In this equation, W_{i} is the penetration loss of internal walls in dB, which is uniformly distributed between 4 dB and 10 dB. Please note that the value of W_{i} is randomly generated for each building and then applied to all of its internal walls in this thesis. In addition, N_{p} represents the number of penetrated internal walls with $N_{\mathrm{p}} \in [0, 1, 2, 3]$, where $N_{\mathrm{p}} = 3$ shows the case that the device locates deeply indoor, e.g. in the basement. In this work, the percentage of devices mapped to the case of $N_{\mathrm{p}} = 3$ is 20%, and the other devices are equally distributed among the cases of $N_{\mathrm{p}} \in [0, 1, 2]$. Moreover, d_{p} and n_{f} stand for the penetration distance in m and the number of penetrated floors, and they are uniformly distributed in the range of $[0, 15]$ and $[0, 4]$, respectively. Last but not least, the external wall penetration loss PL_{ew} in dB also follows a uniform distribution, and the distribution ranges are captured in Table 2.4. Please note that this model will be used later in Section 4.5.

2.3.3.2.4 Indoor-to-indoor propagation model for D2D communication If two sensor devices located in the same building directly communicate

Table 2.4: External wall penetration loss [3GP15a]

PL_{ew}	4-11 dB	11-19 dB	19-23 dB
Percentage of devices uniformly distributed in range	25%	50%	25%

with each other, the respective pathloss models proposed by 3GPP in [3GP13a] are captured in Table 2.5. In this table, PL_{fs} is the free space loss in dB and d_{3D} is the 3D distance between the two devices. Additionally, d_{fl} is the height of one floor and n_{f} stands for the number of penetrated floors that separate the two devices. Both d_{3D} and d_{fl} are in m. Please note that this model will be used later in Section 4.5.

Table 2.5: D2D pathloss models if both devices are indoor [3GP13a]

Scenario	Pathloss model/dB	Shadow fading std/dB
Sensors on the same floor	$PL = PL_{\mathrm{fs}} + \alpha \cdot d_{\mathrm{3D}}, \alpha = 0.5$ dB/m	log-normal $\sigma = 1.5 \cdot \sqrt{d_{\mathrm{3D}}/d_0}$, $d_0 = 1\mathrm{m}$
Sensors on the different floors	$L = A + PL_{\mathrm{fs}} + B\log_{10}(d_{\mathrm{3D}}/d_{\mathrm{fl}})$ $A = \min(n_{\mathrm{f}}L_{\mathrm{fl}}, A_0)$ $B = 25, A_0 = 25\mathrm{dB}, L_{\mathrm{fl}} = 10\mathrm{dB}$	log-normal $\sigma = 4$

2.3.3.2.5 Direct V2X propagation model in urban scenario As shown in Figure 2.4, the pathloss model between two vehicles in a city environment can experience either an LOS or an NLOS propagation condition. In the case of an LOS propagation, the UMi model stated in Equation (2.17) can be applied by properly adjusting the antenna heights of both the Tx and the Rx to h_{UT}, which is the actual height of the user device [MET13b]. However, the UMi model cannot fit well for the NLOS propagation, and thus the model in [MKH11] is used in Section 5.5.4 for calculating the NLOS propagation pathloss in dB, as

$$PL/\mathrm{dB} = 3.75 + 10\log_{10}\left(\left(\frac{d_{\mathrm{t}}^{0.957}}{(X_{\mathrm{t}}W_{\mathrm{r}})^{0.81}}\frac{4\pi d_{\mathrm{r}}}{\lambda}\right)^{2.69}\right). \tag{2.23}$$

In this model, d_{t} and d_{r} represent the distances from the Tx and Rx to the intersection point in m, and X_{t} is the distance from the Tx to the wall in m. In addition, W_{t} and W_{r} are in m and they denote the widths of the streets where the Tx and the Rx locate, respectively. Moreover, the shadow fading in the NLOS case is modeled as a log-normal distribution with a standard deviation of $\sigma = 4.1$ dB. For a better clarification, Figure 2.4 is introduced to graphically show the different terms in Equation (2.23).

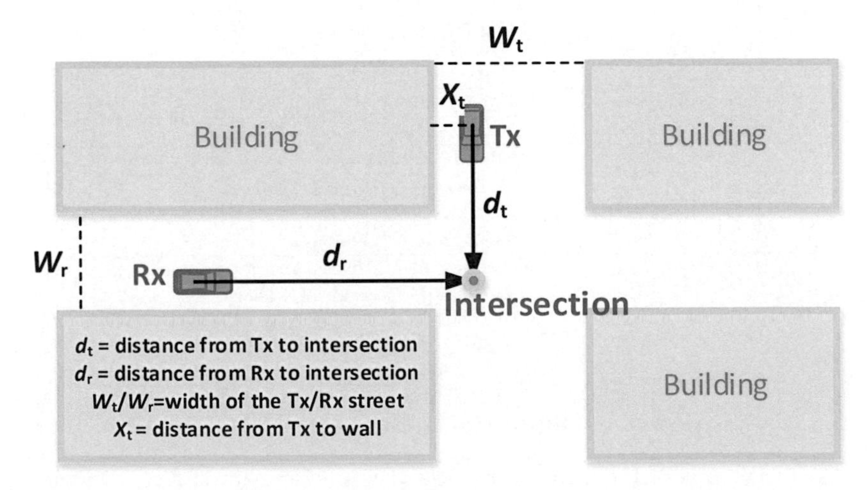

Figure 2.4: Direct V2X channel model

2.3.4 Mobility models

Since D2D communication in this work is applied in different use cases, i.e. eMBB, mMTC, and V2X, the mobility model of UEs should be discussed w.r.t. each specific use case, and detailed information is provided in Table 2.6.

Table 2.6: Mobility models

Use case	UE type	Mobility
eMBB	pedestrians	uniformly distributed in [0,3] km/h
mMTC	sensor devices	stationary
V2X	vehicles	in city: 50 km/h on highway: 120 km/h

2.3.5 Traffic models

As the mobility models, the traffic models in this work are also discussed w.r.t. different use cases and detailed in Table 2.7.

Table 2.7: Traffic models

Use case	UE type	Traffic
eMBB	pedestrians	full buffer [IR08a]
mMTC	sensor devices	urban: Packet size of 250 bytes with a periodicity of 12 packets per hour [JLS17] rural: Packet size of 125 bytes with a periodicity of 24 packets per hour [JHLS17]
V2X	vehicles	packet size of 212 bytes with a periodicity of 10 packets per second [3GP15c]

2.3.6 UE power consumption model

As a critical KPI for mMTC services, the battery life of sensor devices will be evaluated in Section 4.5 according to the power consumption parameters shown in Table 2.8.

2.3.7 Simulation procedure and flow

In Figure 2.5, the simulation flow used in this thesis to generate the cellular network performance is provided. Its details are illustrated below.

1. Set up the scenario for simulation, e.g. choosing between an urban and a rural area. In this step, if a realistic scenario is used, the

Table 2.8: Device power consumption parameters [JHLS17]

Parameter	Description	Value	Time duration if applicable
P^{D}_{tx}	power consumption during transmission	45% PA efficiency plus 60 mW for other circuitry	MCS and packet size related
P^{D}_{rx}	power consumption to receive packets from remote sensors	100 mW	MCS and packet size related
P^{D}_{paging}	power consumption to receive paging command	100 mW	10 ms
P^{D}_{clock}	power consumption to obtain synchronization	100 mW	10 ms
P^{D}_{cp}	power consumption during the CP establishment procedure	200 mW	10 ms
P^{D}_{sleep}	power consumption in sleeping mode	0.01 mW	time of sensor staying in sleeping mode
D_{rx}	number of DRX cycles per day	4 times/day	
t_{DRX}	length of DRX cycle	6 hours	
T_{TMS}	periodicity of TM update	1 day	

detailed geometrical information should be properly modeled in the simulator, e.g. to place buildings and roads.

2. Deploy the network infrastructures in the simulator, e.g. deploy BSs and their equipped antennas. Please note that, in order to avoid margin-effects, a wrap-around model can be applied by deploying additional cells in another tier surrounding the simulation area [SZJ+09]. The simulation area is the area where the performances of all UEs inside this area are collected by the simulator to generate system performance.

3. In each simulation drop, UEs are distributed in the map. The distribution of UEs should be aligned with the considered scenario. For instance, sensors are distributed in a 2D plane in a rural area, while inside 3D buildings in an urban area. Moreover, a drop in

the simulation corresponds to a simulation session that runs for the duration of the drop. Please note, the number of drops should be large enough to ensure a convergence in system performance.

4. Each UE residing in the simulation area is associated with a concrete channel model, e.g. O2I or O2O channel, taking account of the scenario setup in Step 1, e.g. in a rural or an urban area. Afterwards, for a radio link between a UE and the antenna of each cell, its LOS/NLOS channel state and the pathloss value are derived according to the channel models provided in Section 2.3.3. Moreover, the shadow fading factor for each cellular radio link should also be generated based on its standard deviation value. Last but not least, the antenna gains, i.e. transmit antenna gain and receive antenna gain of each link are computed based on the antenna pattern and the geometrical locations of both the transmitter and receiver.

5. Taking account of different parameters, e.g. the previous calculated pathloss value, shadow fading, antenna gains, transmit power of BS, and system operating bandwidth, the received signal power from different cells to a UE can be computed.

6. For each UE residing in the simulation area, attach it to the cell which has the strongest received signal power. Afterwards, the wideband SINR value of each UE in the downlink can be computed by modeling the received signal power from the non-serving cells as interference.

7. Deriving the uplink SINR value for each UE located in the simulation area. The SINR calculation of an uplink should take account of the transmissions of other UEs that take place over the same time-and-frequency resource.

8. Each UE is attached to a traffic class, and its dynamics of data traffic should be generated accordingly.

9. If some UEs have data in their buffers and request radio resources for transmissions, their multi-path parameters, e.g. cluster power, delay, angle of arrival (AoA), angle of departure (AoD), velocity, are used to generate the frequency-selective channel. This information

is later used by the scheduler and the physical layer abstraction approach.

10. Packet transmissions are scheduled by the network with its implemented RRM algorithm. The RRM algorithm can take different metrics into considerations, such as fairness among different UEs and overall system capacity.

11. If a UE is scheduled with a resource for its packet transmission, its radio link performance is generated by applying a link-to-system level mapping approach, e.g. the MIESM approach used in this work, which abstracts and predicts the link layer performance in a computationally simple manner.

12. Performance of the UEs inside the simulation area will be collected and stored.

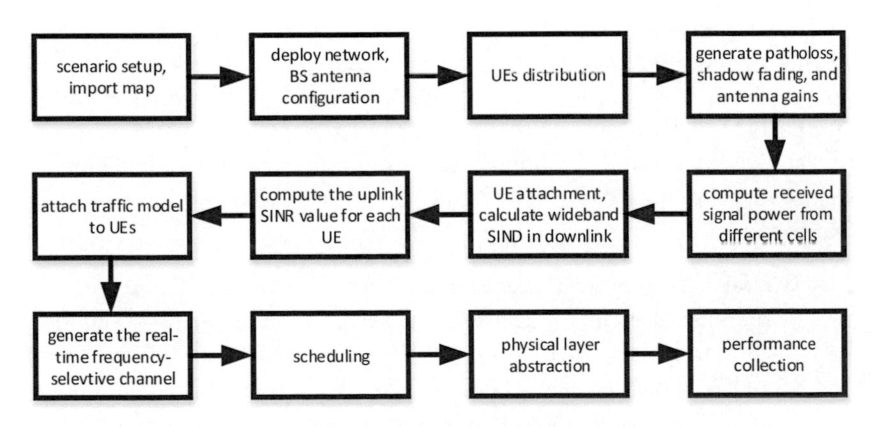

Figure 2.5: Simulation flow

It is worth noticing that some above-calculated information can be pre-calculated and cached in the database before the main simulation loop starts. For instance, if the simulation area has a dimension of M meters $\times N$ meters and a de-correlation distance of ΔD_{cor}, e.g. 5 m is pre-defined, a uniformly spaced grid G with a dimension of $\lceil \frac{M}{\Delta D_{\text{cor}}} + 1 \rceil \times \lceil \frac{N}{\Delta D_{\text{cor}}} + 1 \rceil$ can be constructed. Each element $G_{(m,n)}$ in the grid G represents the received signal power in Step 5 from a specific BS to the geometrical location with a coordinate of $((m-1) \times \Delta D_{\text{cor}}, (n-1) \times \Delta D_{\text{cor}})$.

After computing the value of every element of G, it can be cached in the database. Please note, since there can be multiple cells in the simulation, multiple grids need to be constructed, where each one represents the received signal power from one cell. With this pre-generated information, the simulator can simply compute and associate the received signal power value with the location of a UE by interpolation without the need to calculate the pathloss value and antenna gains of each UE on the fly. Thus, this approach can efficiently reduce the required time of conducting a system-level simulation. Besides the received signal power, other information can also be pre-generated and cached in the database, such as wideband SINR, mobility traces, serving cell attachment at each location, shadow fading, and fast fading traces.

2.4 Summary

In this chapter, the author has pointed out the inadequacies of the legacy evaluation methodology for 4G system. Due to lack a proper modeling on the vertical dimension effects, the legacy 4G evaluation methodology needs to be improved. Thus, in order to capture the properties of realistic scenarios and develop a valid simulation platform, the 5G evaluation methodology described in this chapter compromises the following features:

- Besides the synthetic scenario used for 4G system, a realistic scenario has been introduced and described. This scenario models the environment with more details compared with the synthetic scenario. For instance, the positions and dimensions of buildings, streets, and parks should be taken account in the simulator. As another important factor influencing the simulation performance, the channel model of 4G system is extended to capture the 3D effects. In the extended 3D channel model, the simulator uses a ray tracing method to depict the signal propagation and generates the corresponding pathloss value.

- Different models have been elaborated for different 5G service types, i.e. eMBB, mMTC, and V2X communications. For example, the V2X communication is expected to happen among outdoor users for traffic safety and autonomous driving use cases. In contrast, the mMTC is used for both outdoor and indoor devices to sporadically report their status. Thus, the different 5G services have different features and they need to be appropriately considered in the simulation models, e.g. the traffic model, the mobility model, and user deployment model.

- The service-specific KPIs have been described in detail. Compared with 4G, different 5G services require divergent QoS and face different technical challenges. Therefore, the author has introduced the KPIs used in this thesis for evaluating the technical concepts regarding different service types.

The described models in this chapter are implemented in the simulator, which will be used in the next chapters to generate the simulation results.

Chapter 3
Context-aware Network-controlled Sidelink Communications to Offload Network Traffic

3.1 Introduction

By taking the advantage of geographical proximity, the direct D2D communication is able to improve the performance of cellular networks w.r.t. spectral efficiency, coverage and data rates [FDM+12]. Due to these advantages, some new services can be better supported by D2D communication in 5G compared to 4G, such as direct multimedia transmissions [SKCH11, CLL+10]. In order to obtain a global and optimal coordination among different radio links, the D2D communication in 5G should be controlled by the network operators. In addition, the efficiency of the implemented RRM scheme is closely related with the achievable performance of D2D communication. The topics in RRM mainly cover a transmit power control, a mode selection and a time-and-frequency resource allocation.

In D2D communication, power control is usually considered as an effective mechanism to control interference [FLYW+13, YTDR09, FR11]. On the one hand, the communication range in which the transmitted signals can be successfully recovered at a receiver is determined by the transmit power, since a higher transmit power can cope with a higher radio propagation loss. On the other hand, due to the fact that the same resource element will be reused by different radio links in a cellular network, a higher transmit power will increase the interference power level for the other links transmitting over the same radio resource.

Besides the transmit power control, the operation mode of D2D communication plays also an important role in the RRM. In order to provide ProSe and optimize the hop gain, the spectral resource needs to be efficiently allocated to D2D links. Currently, three different D2D operation modes are envisioned:

- Cellular mode: The point-to-point (P2P) traffic is transmitted via the network infrastructures, as in the traditional cellular network.

- Dedicated mode: A portion of resources will be dedicated by the cellular network to operate the D2D communication.
- Reuse mode: The D2D communication underlaying cellular network will take place over some resources of the cellular links. The spectrum reuse can be either between the sidelink and uplink or between the sidelink and downlink.

In literature, the coexistence of D2D and cellular communications using the same spectral resource, i.e. the reuse mode, is not considered in some work and thus only the dedicated D2D operation mode is studied in [FDM$^+$12, GBCC11, WCC$^+$11]. In this dedicated mode, the D2D communication is realized by simply cutting off some spectral resources from the cellular communication. The advantage of applying the dedicated mode is to obtain an isolation between the ProSe and the other services with low complexity. However, its spectral efficiency is lower than that of the reuse operation mode, since no resource is allowed for being reused. Therefore, in order to improve the efficiency of frequency resources, the reuse mode where D2D and cellular communications share the same resource has been investigated in [FDM$^+$12, WCC$^+$11, YDRT11, BY12, DRW$^+$09]. This mode is essential to provide a large traffic volume when a high spectral efficiency is demanded for the network. However, the D2D reuse mode introduces mutual interference between two links sharing the same resource.
In this chapter, the author investigates the D2D communication in the reuse mode. In order to mitigate the mutual interference and improve the communication efficiency, context information is applied in the RRM, e.g. channel information, UE-specific data rate requirements, service priority, and UE location. In Section 3.2, the network-controlled resource allocation algorithms are proposed to efficiently maximize the system capacity and the spectrum reuse. The proposed algorithms take context information into account, e.g. channel information, and have low computation complexity. As mentioned before, only CSI is used in this section to derive the proposed RRM algorithm as a starting point. Later in Section 3.3, other UE-specific context information, e.g. service data rate requirement and link priority information, will be considered in the smart resource allocation algorithms to improve the quality of experience (QoE). Besides, the author also designs signaling schemes in Section 3.4

to enable the proposed context-aware network-controlled D2D resource allocation schemes. Finally, the chapter is summarized in Section 3.5.

3.2 System capacity optimization for D2D underlay operation

3.2.1 System Model

In this chapter, the overall mobile UEs are divided into two subsets, i.e. D2D UEs and cellular UEs. Two D2D UEs that locate in the geographical proximity of each other can establish a D2D link for ProSe and mutually exchange data. On the other hand, each cellular UE will be served by a BS with one resource block (RB) via uplink and downlink. As the D2D communication in this work operates in the reuse mode, no resource will be dedicated exclusively to the D2D communication. As mentioned before, since the uplink resource is often less utilized and the cellular pilot and synchronization signals are always transmitted in the downlink [LAGR14], a D2D pair can only reuse the RB of a cellular uplink in this chapter. In the legacy cellular networks, an RB is used only by one cellular link in a cell. Thus, it is assumed that an RB cannot be reused by two D2D links in a cell, and the RB of a cellular link can be reused maximally by one D2D pair in this section. Later in Section 3.3, this constraint is relaxed, and it is assumed that two D2D links can also reuse the same RB. As shown in Figure 3.1, in total N cellular users and M D2D pairs are assumed active in a cell where FDD is applied, and $N \leq M$. Please note that only one D2D pair is plotted in this figure for simplicity.

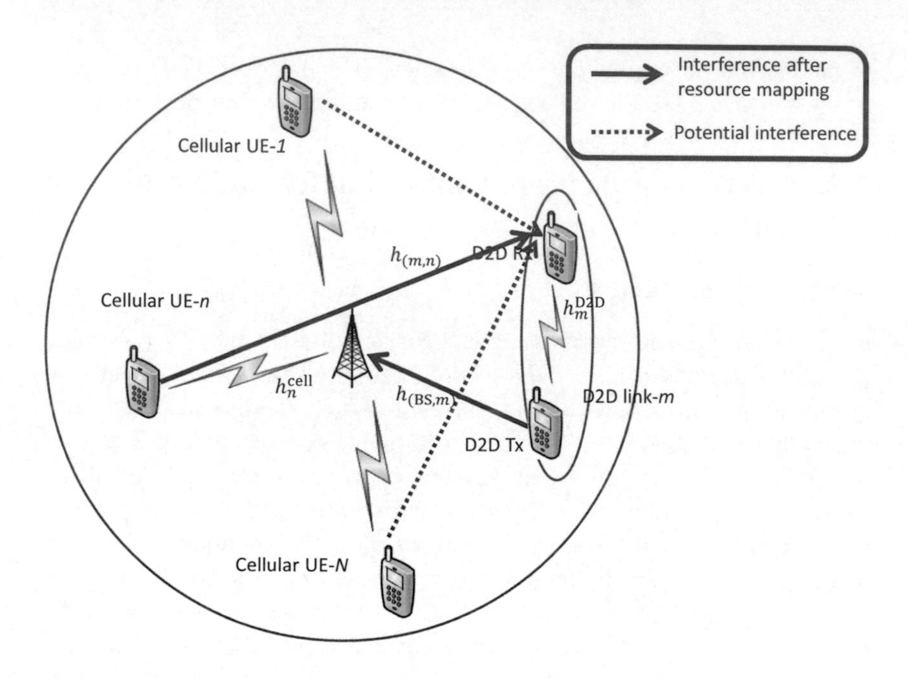

Figure 3.1: Interference scenario: D2D and cellular links use the same resource [JKKS14]

Moreover, the same spectrum band will be shared by the D2D and cellular communications. Thus a resource allocation scheme running at a central network entity, e.g. the BS, tries to assign the same RB to one D2D link and one cellular link. For instance, once the BS receives the scheduling request message from a D2D Tx, it will check if it is feasible to assign the transmission resource of a cellular uplink to the D2D transmission. If yes, the BS reserves the resource and sends the configuration information in downlink to the D2D Tx. Thus, the D2D pair and cellular uplink can transmit over the same resource. In order to optimize the system performance, context information, e.g. channel information, can be exploited to support a smart resource allocation with different aims, e.g. to enable more D2D links or maximize the overall system capacity. Due to the simultaneous data transmissions of the D2D and cellular links over the same resource, a mutual interference will result as shown in Figure 3.1. However, the interference power should be controlled in a way that the quality of both the cellular and D2D links are not severely deteriorated. In other words, in order to assign the same RB to both the mth D2D pair and the nth cellular link, the SINR values of both links, i.e. $SINR^{\text{D2D}}_{(m,n)}$ and $SINR^{\text{cell}}_{(m,n)}$, should remain above the network-controlled target thresholds, i.e. $SINR^{\text{D2D}}_{\text{target}}$ and $SINR^{\text{cell}}_{\text{target}}$, respectively, as

$$SINR^{\text{D2D}}_{(m,n)} \geq SINR^{\text{D2D}}_{\text{target}}, \tag{3.1}$$

and

$$SINR^{\text{cell}}_{(m,n)} \geq SINR^{\text{cell}}_{\text{target}}. \tag{3.2}$$

Please note that the target SINR values should be properly configured so that they can be achieved if there is no mutual interference. Otherwise, though a negligible mutual interference power is introduced, the network will not configure the two links over the same transmission resource. Moreover, the network operator needs to set these target values according to their implementation strategy. For instance, in this section, the target SINR values are set to be -7 dB, since this is the minimal SINR requirement to support a data transmission with the LTE air interface[JF12]. In addition, $SINR^{\text{D2D}}_{(m,n)}$ and $SINR^{\text{cell}}_{(m,n)}$ in Equations (3.1) and (3.2) can be computed as

$$SINR^{\text{D2D}}_{(m,n)} = \frac{h^{\text{D2D}}_{m} P^{\text{D2D}}_{\text{tx}}}{h_{(m,n)} P^{\text{cell}}_{\text{tx}} + \sigma^2_{\text{n}}}, \tag{3.3}$$

and

$$SINR^{\text{cell}}_{(m,n)} = \frac{h^{\text{cell}}_{n} P^{\text{cell}}_{\text{tx}}}{h_{(\text{BS},m)} P^{\text{D2D}}_{\text{tx}} + \sigma^2_{\text{n}}}. \tag{3.4}$$

In Equations (3.3) and (3.4), h^{cell}_{n} denotes the channel gain of the nth cellular link, and h^{D2D}_{m} gives the channel gain between the two transmission ends of the mth D2D pair. In addition, $P^{\text{D2D}}_{\text{tx}}$ and $P^{\text{cell}}_{\text{tx}}$ represent the transmit powers of the D2D and cellular Txs, respectively. The BS can set the transmit power of a link as either a fixed value or a dynamically changed value adapted by the power control mechanism. Thus, the BS can be aware of the transmit power. In this section, the transmit powers of both the D2D and cellular Txs are set to the maximal transmit power of a user terminal, i.e. 24 dBm. Moreover, in order to characterize the interference, $h_{(\text{BS},m)}$ is used to stand for the channel gain between the mth D2D Tx and the BS, and $h_{(m,n)}$ for the channel gain from the nth

cellular UE to the mth D2D Rx. The term σ_{n}^2 denotes the thermal noise power and the noise figure of the Rx. Please note that each channel gain value in Equations (3.3) and (3.4) should be the summation of pathloss value, shadow fading value, and the antenna gain in dB. Moreover, the approach to collect the channel gain information will be discussed later in Section 3.4.

After using the channel gain information to compute $SINR^{\text{D2D}}_{(m,n)}$ and $SINR^{\text{cell}}_{(m,n)}$, the BS is able to run its resource allocation algorithm and derive which cellular RB should be reused by a D2D pair. In general, the resource allocation algorithm will be implemented by the network operators with their own strategies. In this section, the author designs two different algorithms which can optimize the network performance w.r.t. the number of feasible D2D links and the overall system capacity.

3.2.2 Maximize the number of D2D links

In this section, the resource allocation scheme considers the number of feasible D2D links as a KPI and tries to maximize it. A number of M D2D pairs are feasible in a system if the M D2D pairs can reuse the resources of cellular links while exhibiting a good radio quality. This KPI is quite important in application scenarios where a large number of concurrently participating D2D UEs are required, e.g. the public safety use case [3GP14], which is under significant consideration of 3GPP.

3.2.2.1 Construction of the optimization problem

After computing the SINR values by using Equations (3.3) and (3.4), a feasibility matrix can be defined as

$$\boldsymbol{F} = \begin{pmatrix} f(1,1) & \cdots & \cdots & \cdots & f(1,N) \\ \vdots & \ddots & & & \vdots \\ \vdots & & f(m,n) & & \vdots \\ \vdots & & & \ddots & \vdots \\ f(M,1) & \cdots & \cdots & \cdots & f(M,N) \end{pmatrix}, \tag{3.5}$$

which is an $M \times N$ binary matrix. Each element of $f(m,n)$ in this matrix stands for the feasibility function and has a binary value as

$$f(m,n) = \begin{cases} 1, & \text{if both Equations (3.1) and (3.2) are fulfilled;} \\ 0, & \text{else.} \end{cases} \quad (3.6)$$

This feasibility function implies whether the mth D2D pair and the nth cellular user can simultaneously transmit over the same frequency resource. The value of the feasibility function is set to "1" if the resource-reuse between the D2D pair and the cellular UE is feasible and thus one contribution can be made to the number of total feasible D2D links. In the other case, if the value of the feasibility function is "0", the resource-reuse is not feasible and, therefore, it cannot bring any contribution to the number of feasible D2D links. Thus, an optimization problem can be constructed as follows:

$$\textbf{maximize}_{\boldsymbol{X}^{\text{CD}}} \quad \sum_{(m,n)} f(m,n)x^{\text{CD}}(m,n), \quad (3.7)$$

$$\textbf{subject to} \quad x^{\text{CD}}(m,n) \in \{0,1\}, \quad (3.8)$$

$$\sum_{m} x^{\text{CD}}(m,n) \in \{0,1\}, \quad (3.9)$$

$$\sum_{n} x^{\text{CD}}(m,n) \in \{0,1\}, \quad (3.10)$$

$$m \in (1,2,\ldots,M), \quad (3.11)$$

$$n \in (1,2,\ldots,N), \quad (3.12)$$

$$f(m,n) \in \{0,1\}. \quad (3.13)$$

Equation (3.7) has a value, which is smaller or equal to $\min(M,N)$. It shows the objective function to maximize the number of feasible D2D links by searching for the optimal resource allocation matrix $\boldsymbol{X}^{\text{CD}}$, as

$$\boldsymbol{X}^{\text{CD}} = \begin{pmatrix} x^{\text{CD}}(1,1) & \cdots & \cdots & \cdots & x^{\text{CD}}(1,N) \\ \vdots & \ddots & & & \vdots \\ \vdots & & x^{\text{CD}}(m,n) & & \vdots \\ \vdots & & & \ddots & \vdots \\ x^{\text{CD}}(M,1) & \cdots & \cdots & \cdots & x^{\text{CD}}(M,N) \end{pmatrix}, \quad (3.14)$$

where the variable $x^{\text{CD}}(m,n)$ denotes the element on the mth row and the nth column in matrix $\boldsymbol{X}^{\text{CD}}$. And $x^{\text{CD}}(m,n)$ has a binary value as

defined in Equation (3.8). If the resource allocation scheme assigns a value of "1" to $x^{\mathrm{CD}}(m,n)$, it indicates that the mth D2D pair reuses the RB of the nth cellular user. Moreover, please note that the matrix $\boldsymbol{F}$ shows which cellular resources could be principally reused by D2D links, and the matrix $\boldsymbol{X}^{\mathrm{CD}}$ shows which of the cellular resources are actually reused. In addition, the number of elements "1" in each row of the matrix $\boldsymbol{F}$ is in the range $\{0, N\}$. In comparison, the element of "1" in each row of the matrix $\boldsymbol{X}^{\mathrm{CD}}$ cannot appear more than once, because each cellular resource shall not be reused by more than one D2D link.
In addition, Equations (3.9) to (3.12) show the constraints of the resource allocation problem that the resource of each cellular user can be maximally reused by one D2D pair, and each D2D pair can at most reuse the RB of one cellular user.
As it can be seen, the above optimization problem is an integer linear programming problem. Thus, to compute the optimal solution requires a high computational complexity at the central entity. For example, for an exhaustive search for the optimal solution, there are in total $P_{\max(M,N)}^{\min(M,N)}$ resource allocation options to be computed, where P denotes the permutation operation and $P_n^m = \frac{n!}{(n-m)!}$.

3.2.2.2 Heuristic RRM algorithm

As aforementioned, in order to check the feasibility of assigning the same resource to both a cellular uplink and a D2D link, it is necessary for the BS to collect the necessary channel gain information, including the channel gains between a D2D Tx and its Rx, a cellular UE and the BS, between each cellular UE and a D2D Rx, and between each D2D Tx and the BS. The first two channel information terms are used to calculate the received signal strengths for the D2D link and the cellular link, respectively, and the last two are required to compute the interference power for these two links.
Once the network obtains an awareness of the overall channel information, it can run its resource allocation algorithm and search for the solution of the optimization problem shown in Equations (3.7) to (3.13). However, as pointed out before, a high computational complexity is required to find the optimal solution for the integer linear programming problem. In this section, a heuristic resource allocation algorithm based

on the branch-and-cut algorithm [PR91] is introduced as Algorithm 1 that is able to solve the proposed optimization problem while exhibiting a modest complexity. Please note that the algorithms introduced in this thesis are implemented by using MATLAB. For an easy understanding, a simple example has been introduced in Figure 3.2.

In Algorithm 1, the key idea is to give a high priority to the users $\hat{m}$ which have the least feasible options. In order to maximize the total number of D2D links, a D2D link should try to use the cellular resource not feasible by other D2D links. As defined in Equation (3.6), if $m = \hat{m}$, the different elements $f(m,n)$ in the $\hat{m}^{\text{th}}$ row of matrix $\boldsymbol{F}$ show the feasibility to assign the resources of different cellular UEs to the $\hat{m}^{\text{th}}$ D2D pair. Thus, the $\hat{m}^{\text{th}}$ row in the matrix $\boldsymbol{F}$ with the least "1"s depicts that the $\hat{m}^{\text{th}}$ D2D pair has the least resource-reuse options, and the resource allocation scheme will try to first assign a resource to this D2D pair. In addition, the resources $\hat{n}$ which can be assigned to the least optional D2D links are also given a high priority in this algorithm. To maximize the total number of D2D links, the cellular RBs should be reused as much as possible. In Equation (3.6), the $\hat{n}^{\text{th}}$ column of matrix $\boldsymbol{F}$ shows the feasibility to assign the resource $\hat{n}$ to the different D2D links. Thus, the $\hat{n}^{\text{th}}$ column with the least "1"s represents that the $\hat{n}^{\text{th}}$ resource can be assigned to the least D2D links and, therefore, the resource allocation scheme will try to first assign this resource to a D2D pair. As we can see from Algorithm 1 and Figure 3.2, in each iteration from Step 7 to Step 22 in Algorithm 1, the resource of one specific cellular UE will be assigned to a D2D pair if there is any feasible assignment option regarding this cellular uplink resource. To be more specific, if a D2D pair has only one feasible resource to reuse or a resource can only be reused by one D2D pair, the algorithm performs the resource allocation in Steps 7 to 8. Otherwise, the algorithm scans for the resource or the D2D link, which has the least feasible options in Step 17, and then assigns the resource in Step 18. After the assignment of each resource, the matrix $\boldsymbol{F}$ will be updated. In detail, all the elements in the $\hat{m}^{\text{th}}$ row and in the $\hat{n}^{\text{th}}$ column will be set to zero, if the resource allocation scheme assigns the $\hat{n}^{\text{th}}$ resource to the $\hat{m}^{\text{th}}$ D2D link, as shown in Steps 10 and 20. Therefore, the algorithm will not consider this resource and this D2D link in the next round. In other words, the resource and the D2D link are cut away from the resource allocation scheme. As we can see from the analysis, the proposed heuristic algorithm has a complexity of $O(N \cdot M)$ to assign a cellular uplink resource to a D2D pair. Once the resource allocation is done, Step 24 to Step 26 will run to generate the resource allocation matrix $\boldsymbol{X}^{\mathrm{CD}}$.

The output of Algorithm 1, $D2DpairNum$ and $CellularNum$ are two vectors used to illustrate the decision of the resource mapping. Another output of Algorithm 1 is the resource allocation matrix $\boldsymbol{X}^{\mathrm{CD}}$, which is derived from vectors $D2DpairNum$ and $CellularNum$. For example, $D2DpairNum(i) = m$ and $CellularNum(i) = n$ indicate that the mth D2D pair is set to reuse the resource of the nth cellular user. Thus, the length of one of these two vectors actually provides the number of overall feasible D2D links achieved by the resource allocation algorithm.

$$\boldsymbol{F} = \begin{pmatrix} 1 & 1 & 0 & 0 & 0 \\ 0 & 1 & 0 & 0 & 1 \\ 0 & 1 & 0 & 1 & 0 \\ 0 & 1 & 0 & 1 & 0 \\ 1 & 1 & 1 & 0 & 1 \end{pmatrix}, \Rightarrow \boldsymbol{F} = \begin{pmatrix} 1 & 1 & 0 & 0 & 0 \\ 0 & 1 & 0 & 0 & 1 \\ 0 & 1 & 0 & 1 & 0 \\ 0 & 1 & 0 & 1 & 0 \\ 0 & 0 & 0 & 0 & 0 \end{pmatrix}, \Rightarrow \boldsymbol{F} = \begin{pmatrix} 0 & 0 & 0 & 0 & 0 \\ 0 & 1 & 0 & 0 & 1 \\ 0 & 1 & 0 & 1 & 0 \\ 0 & 1 & 0 & 1 & 0 \\ 0 & 0 & 0 & 0 & 0 \end{pmatrix},$$

CellularNum: 3
D2DpairNum: 5

CellularNum: 3 1
D2DpairNum: 5 1

$$\Rightarrow \boldsymbol{F} = \begin{pmatrix} 0 & 0 & 0 & 0 & 0 \\ 0 & 0 & 0 & 0 & 0 \\ 0 & 1 & 0 & 1 & 0 \\ 0 & 1 & 0 & 1 & 0 \\ 0 & 0 & 0 & 0 & 0 \end{pmatrix}, \Rightarrow \boldsymbol{F} = \begin{pmatrix} 0 & 0 & 0 & 0 & 0 \\ 0 & 0 & 0 & 0 & 0 \\ 0 & 0 & 0 & 0 & 0 \\ 0 & 0 & 0 & 1 & 0 \\ 0 & 0 & 0 & 0 & 0 \end{pmatrix}.$$

CellularNum: 3 1 5 2
D2DpairNum: 5 1 2 3

CellularNum: 3 1 5 2 4
D2DpairNum: 5 1 2 3 4

$$\boldsymbol{X}^{\mathrm{CD}} = \begin{pmatrix} 1 & 0 & 0 & 0 & 0 \\ 0 & 0 & 0 & 0 & 1 \\ 0 & 1 & 0 & 0 & 0 \\ 0 & 0 & 0 & 1 & 0 \\ 0 & 0 & 1 & 0 & 0 \end{pmatrix}.$$

Figure 3.2: A simple example to illustrate Algorithm 1

Algorithm 1 Maximization of underlaying D2D links [JKKS14]

Input:
The feasibility matrix $\boldsymbol{F}$;
Output:
ID of the combined D2D pairs: $D2DpairNum$;
ID of the combined cellular users: $CellularNum$;
resource allocation matrix $\boldsymbol{X}^{\mathrm{CD}}$;

```
1: initialize an all-zero resource allocation matrix X^CD with dimension of M × N;
2: if all elements in matrix F equal to "0" then
3:   D2DpairNum=[], CellularNum=[];
4:   end the algorithm without assigning any underlaying D2D pair;
5: else
6:   i=1;
7:   if one column or one row vector in matrix F has a unique "1" element at position
     f(m̂, n̂) then
8:     allocate the m̂^th D2D pair with the resource block of n̂^th cellular user:
       D2DpairNum(i) = m̂, CellularNum(i) = n̂;
9:     i=i+1;
10:    set all the elements in the m̂^th row and n̂^th column in matrix F to "0";
11:    if all elements in matrix F equal to "0" then
12:      end the algorithm;
13:    else
14:      return to Step 7;
15:    end if
16:  else
17:    scan all the non zero columns and rows to find the column or row with the
       least number of "1"s, pick out the first position where f(m̂, n̂) = 1 in this
       column or row;
18:    allocate the m̂^th D2D pair with the resource block of n̂^th cellular user:
       D2DpairNum(i)=m̂, CelluluarNum(i)=n̂,;
19:    i=i+1;
20:    set all the elements in the m̂^th row and n̂^th column in matrix F to "0";
21:    return to Step 7;
22:  end if
23: end if
24: for each w ∈ [1, i − 1] do
25:   set the element x^CD(D2DpairNum(w), CellularNum(w)) = 1 in the resource
      allocation matrix X^CD;
26: end for
```

3.2.2.3 Effectiveness of the algorithm

Please note that the heuristic Algorithm 1 offers a sub-optimal solution. As shown in Steps 17 and 18, a resource will be assigned to both D2D and cellular links in a first-in-first-out manner. Therefore, more than

one solution may be possible, and it is necessary to prove the validity of the proposed algorithm before applying it in reality. Without loss of generality, a special case can be considered where the same number of D2D pairs and cellular UEs are deployed in a cell, i.e. $M = N$. From the mathematical perspective, a matrix $\boldsymbol{F}$ can be manually constructed which has the property that all resources can be reused by applying an optimal resource allocation solution. In other words, if this matrix $\boldsymbol{F}$ is fed to the input of a resource allocation algorithm, as an optimal solution all resource blocks will be reused by the D2D pairs.
In order to manually construct the matrix $\boldsymbol{F}$, the following steps should be performed:

1. An identity matrix $\boldsymbol{F}'$ needs to be constructed at first, which has the value of "1"s in its main diagonal while "0"s elsewhere, as

$$\boldsymbol{F}' = \begin{pmatrix} 1 & 0 & \cdots & 0 & 0 \\ 0 & 1 & 0 & \ddots & \vdots \\ \vdots & 0 & \ddots & 0 & \vdots \\ \vdots & \ddots & 0 & 1 & 0 \\ 0 & \cdots & \cdots & 0 & 1 \end{pmatrix}. \tag{3.15}$$

   The identity matrix $\boldsymbol{F}'$ ensures that every RB can be reused theoretically.

2. Afterwards, a random procedure is used to assign the values for the non-diagonal elements in the matrix $\boldsymbol{F}'$, i.e. $f'(m,n)$ where $m \neq n$. Each non-diagonal element is assigned the value of either "0" or "1". In particular, each element $f'(m,n)$ with $m \neq n$ obtains the value of "0" with probability $Prob$ and the value of "1" with probability $(1-Prob)$ from the random procedure. In this approach, the newly generated matrix $\boldsymbol{F}'$ will not reduce the maximal number of the reusable resources, but only provides more solution options for reusing all the resources.

3. As the last step, the matrix $\boldsymbol{F}'$ is randomly interleaved either column-wise or row-wise to yield the matrix $\boldsymbol{F}$. The row-wise interleaving means to randomly pick up two rows of matrix $\boldsymbol{F}'$ and exchange their positions. In contrast, the column-wise interleaving randomly picks up two columns of matrix $\boldsymbol{F}'$ and exchanges their positions.

After that, the interleaved matrix $\boldsymbol{F}$ will be fed to the input of the resource allocation scheme.

In addition to the above mathematical approach, another alternative to evaluate the efficiency of the proposed RRM algorithm is to manually deploy the D2D pairs and the cellular UEs in a simulation environment. In this case, the author intentionally and coordinately deploys a D2D pair and a cellular UE in a way that the channel conditions shown in Equation (3.6) can be fulfilled. With such a deployment, all cellular resources can be reused by D2D links.
Later in Section 3.2.4, the author will show that the proposed sub-optimal algorithm permits to reuse all cellular uplink RBs with the above-mentioned two approaches. Therefore, the effectiveness of the algorithm can be proven.

3.2.3 System capacity maximization

Previously, Algorithm 1 has been developed where the author attempted to maximize the number of feasible D2D links. This KPI is of importance since it is favored in cases to support as many D2D links as possible with a fixed amount of cellular uplink resources. On the other hand, network capacity in some commercial use cases [3GP14] is also critical, since a network operator usually charges its mobile users based on their consumed data volume. Thus, it is imperative to develop a resource allocation scheme which can maximize system capacity.
From the capacity perspective, the network can always assign a cellular resource to the D2D pair that is able to yield the maximal link capacity, no matter how many resources this D2D pair has already occupied. For instance, if the two ends of a D2D link are located very close to each other, a low pathloss will be experienced. Thus, a high capacity can be achieved by this D2D link, and the network can assign a large number of resources to this D2D link. Although this scheme can offer a simple and straightforward solution and obtain the maximal capacity performance, it has certain drawbacks. First of all, due to the spatial distribution of D2D pairs, the D2D Txs located near the BS will generate a higher interference level to the cellular uplink transmission. Therefore, a much lower capacity of a cellular link can be expected if its resource is reused by the D2D pair near the BS, comparing to the case where the cellular

resource is reused by a D2D pair at the cell border. Thus, the D2D pairs near the BS have a lower probability to be assigned any cellular resource, which causes unfairness in reality. Secondly, the D2D pairs assigned with multiple RBs need to operate on discontinuous RBs, and this increases the hardware complexity. In order to provide scheduling fairness and keep hardware complexity low, it is assumed that one D2D pair can only reuse the RB of one cellular link, as already mentioned in Section 3.2.2.1. Thus, a capacity matrix $\boldsymbol{C}^{\mathrm{CD}}$ can be constructed as the input to the RRM algorithm:

$$\boldsymbol{C}^{\mathrm{CD}} = \begin{pmatrix} c^{\mathrm{CD}}(1,1) & \cdots & \cdots & \cdots & c^{\mathrm{CD}}(1,N) \\ \vdots & \ddots & & & \vdots \\ \vdots & & c^{\mathrm{CD}}(m,n) & & \vdots \\ \vdots & & & \ddots & \vdots \\ c^{\mathrm{CD}}(M,1) & \cdots & \cdots & \cdots & c^{\mathrm{CD}}(M,N) \end{pmatrix}, \tag{3.16}$$

where $c^{\mathrm{CD}}(m,n)$ represents the achievable capacity, if the resource of the nth cellular UE is reused by the mth D2D pair. The achievable capacity $c^{\mathrm{CD}}(m,n)$ has a non-negative value and it will be derived based on the received signal quality, i.e. SINR, in the simulator. In this chapter, the mapping table from an SINR value to its capacity shown in [JF12] is exploited to derive the capacity.

Correspondingly, the capacity optimization problem can be constructed as follows:

$$\begin{aligned} \textbf{maximize}_{\boldsymbol{X}^{\mathrm{CD}}} \quad & \sum_{(m,n)} c^{\mathrm{CD}}(m,n) x^{\mathrm{CD}}(m,n), && (3.17) \\ \textbf{subject to} \quad & x^{\mathrm{CD}}(m,n) \in \{0,1\}, && (3.18) \\ & \sum_{m} x^{\mathrm{CD}}(m,n) \in \{0,1\}, && (3.19) \\ & \sum_{n} x^{\mathrm{CD}}(m,n) \in \{0,1\}, && (3.20) \\ & m \in (1,2,\ldots,M), && (3.21) \\ & n \in (1,2,\ldots,N). && (3.22) \end{aligned}$$

Equation (3.17) is the objective function where $c^{\mathrm{CD}}(m,n)$ represents the achievable capacity if the resource of the nth cellular UE is reused by the mth D2D pair. The same constraints in Equations (3.18) to (3.22)

are applied here as in Section 3.2.2.1 to indicate that the corresponding resource allocation is an integer linear programming problem. In each row and each column of the matrix $\boldsymbol{X}^{\mathrm{CD}}$ at most one non-zero element $x^{\mathrm{CD}}(m,n)$ can appear.
It is worth noticing that the objective function in Equation (3.17) can be used to maximize different capacity terms, e.g. the capacity of D2D links, the capacity of cellular links or the capacity sum of both D2D and cellular links. Please also note, if a link has an SINR value lower than the minimal SINR threshold to set up a data transmission, this link is not feasible to reuse the same resource of the other link and it has a capacity of zero. Therefore, the value of each component $c^{\mathrm{CD}}(m,n)$ should be derived by mapping the estimated SINR values, i.e. computed in Equations (3.3) and (3.4), to the corresponding link capacities. Actually, the above optimization problem can be considered as an auctioning problem. Regarding the nth cellular resource block, $c^{\mathrm{CD}}(m,n)$ can be taken as the benefit for assigning this resource block to the mth D2D pair. Thus, the auctioning algorithm in [Ber92] is applied in Algorithm 2 to search for an optimal resource allocation solution. For a better understanding, a simple example is provided in Figure 3.3 for illustrating Algorithm 2. As it can be seen from the example, Algorithm 2 applies an iterative approach to search for the optimal bidder of each object and the exact solution of the optimization problem. Please note that Algorithm 2 can be applied in the case $M \leq N$, i.e. the number of cellular links should be larger than or equal to the number of D2D links.

Input:

$\boldsymbol{C}^{\mathrm{CD}} = \begin{pmatrix} 7 & 8 & 9 \\ 8 & 6 & 5 \\ 1 & 6 & 6 \end{pmatrix}$;

$M = 3$;

Initialization:

$E = (0 \quad 0 \quad 0)$;

$CellularNum = (1 \quad 2 \quad 3)$;

$D2DpairNum = (1 \quad 2 \quad 3)$;

Enter the first for-loop:

m=1;

$\boldsymbol{C}^{\mathrm{CD}}(m,:) - E = (7\ 8\ 9) - (0\ 0\ 0) = (7\ 8\ 9)$;

$n = \arg\max(7\ 8\ 9) = 3$;

$m' = \arg(\mathrm{CellularNum} = n) = 3$;

$CellularNum(m') = CellularNum(m)$

$\Rightarrow CellularNum = (1 \quad 2 \quad 1)$;

$CellularNum(m) = n$

$\Rightarrow CellularNum = (3 \quad 2 \quad 1)$;

$\alpha = 9 - 8 = 1$;

$E_n = E_n + \gamma = 1; \Rightarrow E = (0\ 0\ 1)$;

Increase the value of m in the first for-loop:

m=2;

$\boldsymbol{C}^{\mathrm{CD}}(m,:) - E = (8\ 6\ 5) - (0\ 0\ 1) = (8\ 6\ 4)$;

$n = \arg\max(8\ 6\ 4) = 1$;

$m' = \arg(CellularNum = n) = 3$;

$CellularNum(m') = CellularNum(m)$

$\Rightarrow CellularNum = (3 \quad 2 \quad 2)$;

$CellularNum(m) = n$

$\Rightarrow CellularNum = (3 \quad 1 \quad 2)$;

$\alpha = 8 - 6 = 2$;

$E_n = E_n + \gamma = 2; \Rightarrow E = (2\ 0\ 1)$;

Increase the value of m in the first for-loop:

m=3;

$\boldsymbol{C}^{\mathrm{CD}}(m,:) - E = (1\ 6\ 6) - (2\ 0\ 1) = (-1\ 6\ 5)$;

$n = \arg\max(-1\ 6\ 5) = 2$;

$m' = \arg(CellularNum = n) = 3$;

$CellularNum(m') = CellularNum(m)$

$\Rightarrow CellularNum = (3 \quad 1 \quad 2)$;

$CellularNum(m) = n$

$\Rightarrow CellularNum = (3 \quad 1 \quad 2)$;

$\alpha = 6 - 5 = 1$;

$E_n = E_n + \gamma = 1; \Rightarrow E = (2\ 1\ 1)$;

The first for-loop ends and enter the second for-loop:

m=1;

$\boldsymbol{C}^{\mathrm{CD}}(m,:) - E = (7\ 8\ 9) - (2\ 1\ 1) = (5\ 7\ 8)$;

max(5 7 8)=8;

✓ the element at the pisition of $CellularNum(m)$ in vector (5 7 8) has a value of 8, which is equal to the maximal value;

m=2;

$\boldsymbol{C}^{\mathrm{CD}}(m,:) - E = (8\ 6\ 5) - (2\ 1\ 1) = (6\ 5\ 4)$;

max(6 5 4)=6;

✓ the element at the pisition of $CellularNum(m)$ in vector (6 5 4) has a value of 6, which is equal to the maximal value;

m=3;

$\boldsymbol{C}^{\mathrm{CD}}(m,:) - E = (1\ 6\ 6) - (2\ 1\ 1) = (-1\ 5\ 5)$;

max(-1 5 5)=5;

✓ the element at the pisition of $CellularNum(m)$ in vector (-1 5 5) has a value of 5, which is equal to the maximal value;

Algorithm ends

$CellularNum = (3\ 1\ 2)$,

$D2DpairNum = (1\ 2\ 3)$.

$\boldsymbol{X}^{\mathrm{CD}} = \begin{pmatrix} 0 & 0 & 1 \\ 1 & 0 & 0 \\ 0 & 1 & 0 \end{pmatrix}$.

Figure 3.3: A simple example to illustrate Algorithm 2

Algorithm 2 Auction algorithm for the maximization of system capacity [JKKS14]

Input:
The capacity matrix $\boldsymbol{C}^{\mathrm{CD}}$;
Output:
ID of the combined D2D pairs: $D2DpairNum$;
ID of the combined cellular users: $CellularNum$;
resource allocation matrix $\boldsymbol{X}^{\mathrm{CD}}$;

1: initialize an all-zero resource allocation matrix $\boldsymbol{X}^{\mathrm{CD}}$ with dimension of $M \times N$;
2: initialize an all-zero vector E with dimension of $1 \times N$;
3: **for** each $n \in [1, N]$ **do**
4: set $CellularNum(n)= n$;
5: **end for**
6: **for** each $m \in [1, M]$ **do**
7: set $D2DpairNum(m) = m$;
8: **end for**
9: **for** each $m \in [1, M]$ **do**
10: subtract vector E from the mth row of matrix $\boldsymbol{C}^{\mathrm{CD}}$;
11: find an object resource block n which offers the maximal value in the result of Step 10;
12: find element index of m' where $CellularNum(m')=n$;
13: set $CellularNum(m')=CellularNum(m)$;
14: set $CellularNum(m)=n$;
15: set $\alpha = v - w$, where v and w are the maximal and second maximal values in the result of Step 10, respectively;
16: set E_n, which is the nth element in vector E, as $E_n = E_n + \alpha$;
17: **end for**
18: **for** each $m \in [1, M]$ **do**
19: subtract vector E from the mth row of matrix $\boldsymbol{C}^{\mathrm{CD}}$, i.e. $\boldsymbol{C}^{\mathrm{CD}}(m, :)$;
20: **if** the maximal value of the above result is not equal to the value in the position of $CellularNum(m)$ **then**
21: return to Step 9;
22: **end if**
23: **end for**
24: **for** each $m \in [1, M]$ **do**
25: set the element $x^{\mathrm{CD}}(D2DpairNum(m), CellularNum(m)) = 1$ in the resource allocation matrix $\boldsymbol{X}^{\mathrm{CD}}$;
26: **end for**

At the beginning of Algorithm 2, a price vector E is initialized to an all-zero vector, and it indicates the price that each D2D link needs to pay in order to be assigned the different resource blocks. The value of each element in E will be updated by the algorithm when the algorithm runs. In each iteration from Step 9 to Step 17, a cellular resource will be allocated to a D2D link. In Step 11, the algorithm searches for the nth

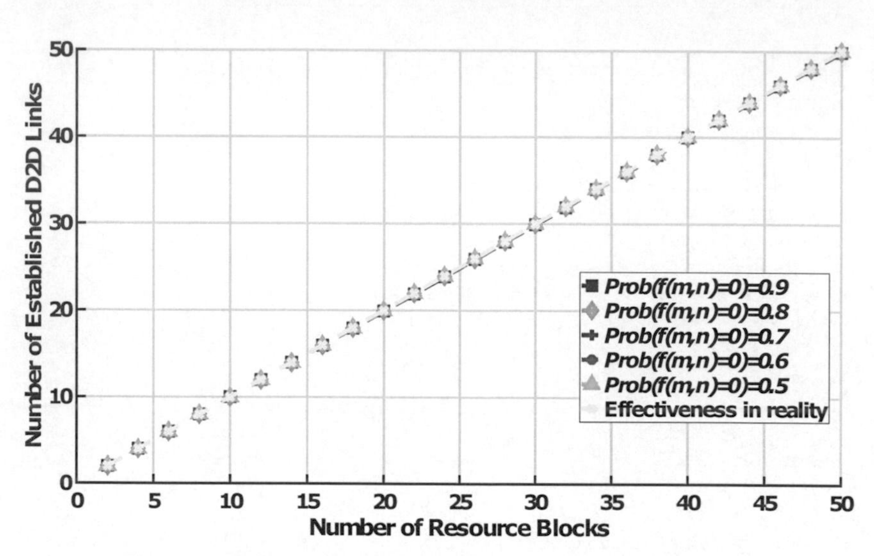

Figure 3.4: Effectiveness demonstration of Algorithm 1 [JKKS14]

resource block which offers the maximal capacity gain to the considered m^{th} D2D link. After that, in Step 12 to Step 14 the mth D2D link exchanges the resource block with the D2D link assigned with the nth resource block at the beginning of the round, i.e. the D2D link m' in Algorithm 2. In Step 15 and Step 16 the price vector E will be updated, and it will be used in Step 10 in the next round to calculate the gain by allocating a resource to the next D2D link. Once the loop from Step 9 to Step 17 has been executed, the algorithm enters the other loop from Step 18 to Step 23, which is used to check whether an optimal solution has been generated or not. If not, the algorithm returns to Step 9, and the iteration from Step 9 to Step 17 will run again. If yes, the resource allocation matrix $\boldsymbol{X}^{\text{CD}}$ will be generated from Step 24 to Step 26. As the outputs of the Algorithm 2, $D2DpairNum$ and $CellularNum$ illustrate the resource mapping decision, and the matrix $\boldsymbol{X}^{\text{CD}}$ is the resource allocation matrix.

3.2.4 Simulation parameters and numerical results

In Figure 3.4, the author first shows the effectiveness of the proposed resource allocation Algorithm 1 that maximizes the number of established

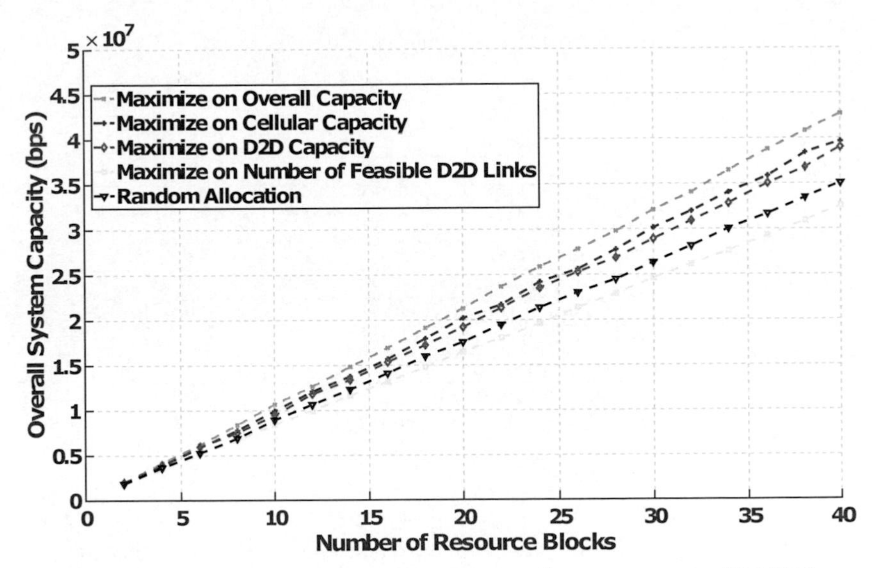

Figure 3.5: Performance comparison w.r.t. overall system capacity [JKKS14]

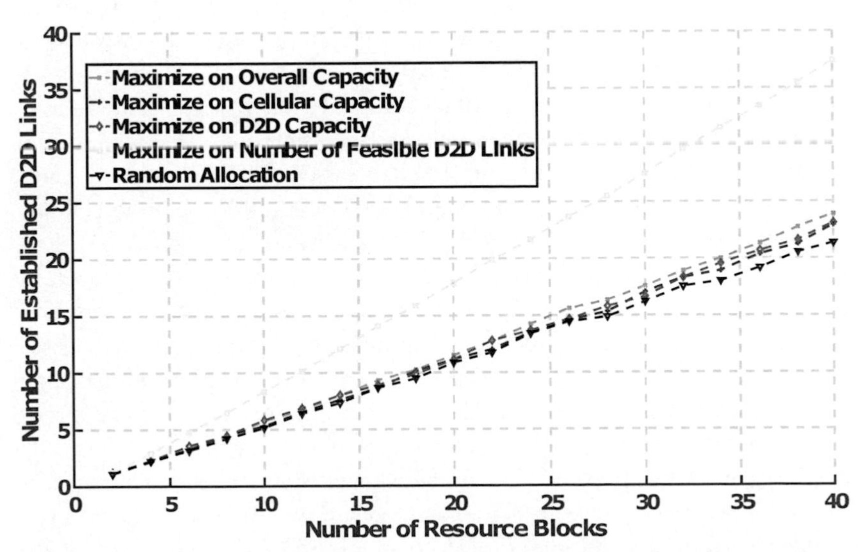

Figure 3.6: Performance comparison w.r.t. number of established D2D links [JKKS14]

Table 3.1: Simulation parameters [JKKS14, IR08a]

Parameter	Value
Cell radius	500 m
Noise figure at BS	5 dB
Noise figure at D2D Rx	7 dB
Target cellular SINR value	-7 dB
Target D2D SINR value	-7 dB
Power level of thermal noise	-174 dBm/Hz
Cable loss	3 dB
Bandwidth per cellular user	180 KHz
Transmitting power of both cellular UE and D2D Tx	24 dBm
Antenna patterns for all devices	omni-directional
SINR estimation	perfect
Maximal feasible distance between one D2D pair	120 m

D2D links in the case $M = N$. As the first step, an identity matrix will be formed, which means that the resource of the nth cellular uplink can be reused by the nth D2D link. Afterwards, binary values will be allocated to the non-main diagonal elements $\{f(m,n), m \neq n\}$ based on a stochastic process, and then the matrix will be randomly interleaved either column-wise or row-wise in order to construct the feasibility matrix $\boldsymbol{F}$ of Equation (3.5). Since it is assumed that to each cellular user is assigned only one RB, the number of cellular links is equal to the number of RBs. The obtained matrix $\boldsymbol{F}$ will be fed to the input of Algorithm 1, and an optimal solution should enable a full reuse of all cellular resources, as stated in Section 3.2.2.3. In addition, the practical deployment scenario shown in Section 3.2.2.3, where the users are deployed in a way that all cellular resources are reusable, is also used to validate the resource allocation proposal. The curves in Figure 3.4 demonstrate that all RBs can be reused by the proposed resource allocation scheme without any noticeable loss. As shown in this figure, the five curves representing different probabilities of matrix elements with a value of "0" overlap with the curve, which represents the practical deployment scenario. This observation can demonstrate that the proposed Algorithm 1 is able to achieve the optimal objective under different conditions. Please note that the matrix $\boldsymbol{F}$ generated manually is only used to validate the proposed Algorithm 1 in Figure 3.4. In the following simulation results, $\boldsymbol{F}$ will be generated based on the real simulation scenario.

In order to derive the performance of the proposed schemes, a system-level simulator is implemented using the synthetic model introduced in Section 2.3. Moreover, the simulation assumptions are aligned with the 3GPP simulation baseline document TR 36.814 [3GP10], and the corresponding parameters are captured in Table 3.1. These parameters are used to set up the simulator. As mentioned in Section 2.3, the UMa channel models proposed in [3GP13b, 3GP10] are referred to as the baseline for channel models of D2D and cellular links correspondingly. In addition, the author also refers to [JF12] for the mapping table from SINR values to capacity values for both the D2D link and the cellular uplink. Last but not least, the distance between the Tx and the Rx of a D2D pair is uniformly and randomly distributed within 120 m. These assumptions and parameters are used to generate the matrices $\boldsymbol{F}$ and $\boldsymbol{C}^{\mathrm{CD}}$ in the simulator.
In Figure 3.5, the system capacity in five different schemes is provided, as listed below:

- Maximization on Overall Capacity: $c^{\mathrm{CD}}(m,n)$ in Equation (3.17) represents the summation of the D2D capacity and the cellular capacity.
- Maximization on Cellular Capacity: $c^{\mathrm{CD}}(m,n)$ in Equation (3.17) represents the cellular capacity.
- Maximization on D2D Capacity: $c^{\mathrm{CD}}(m,n)$ in Equation (3.17) represents the D2D capacity.
- Maximization on Number of Feasible D2D Links: Algorithm 1 is applied.
- Random Allocation: for each D2D pair, a random resource allocation is applied.

As can be seen from Figure 3.5, the three schemes trying to maximize the three different capacity terms outperform the random allocation scheme. More specifically, since the overall capacity of both the D2D and cellular links is maximized in Algorithm 2, it has a performance gain of 20% compared with the random allocation scheme. However, the performance of Algorithm 1 has even a worse performance regarding system capacity compared with the random case. This is because a high priority is given to the set of users which have less feasible options in Algorithm 1. Since these users experience low SINR values, they are more vulnerable. Correspondingly, if these users reuse the cellular resource, the achievable capacity is lower than in the average case and, thus, it can explain the observation why Algorithm 1 has the worst performance in Figure 3.5. In Figure 3.6, the different schemes are inspected w.r.t. the number of established D2D links. Since the RRM algorithm gives a high priority to vulnerable users by first assigning available resources to them, Algorithm 1 has a better performance gain than the other schemes. Compared with the random allocation scheme, it provides a performance gain of 68% approximately. In addition, the schemes maximizing different capacity terms yield a slightly better performance compared with the random allocation scheme.

3.3 Dynamic context-aware optimization of D2D w.r.t. user satisfaction

In 5G, context information needs to be identified from different sources, e.g. a UE and a BS, and it will act as a key to support an efficient RRM [MET17c]. Under the scope of IMT-2020, context-awareness is defined as to deliver real-time context information of the network, devices, applications, the user and its environment to applications and network layers [IR14]. The gathered context information can be exploited by the resource allocation algorithm. In this section, the author develops a context-aware resource allocation scheme which takes account of the user-specific information, e.g. priority and service data rate requirement.

3.3.1 System model

In this section, an RB cannot be simultaneously reused by two different cellular uplinks inside a cell. However, it can be reused by two different D2D links, as long as the mutual interference is weak enough. Therefore, in a cell where many cellular UEs are active, a resource-reuse can only take place between a cellular uplink and a D2D pair. While in another cell with less cellular UEs but many D2D pairs, part of the total resources can be reused between a cellular uplink and a D2D pair while the other resources can be used to support resource-reuse between two D2D links.

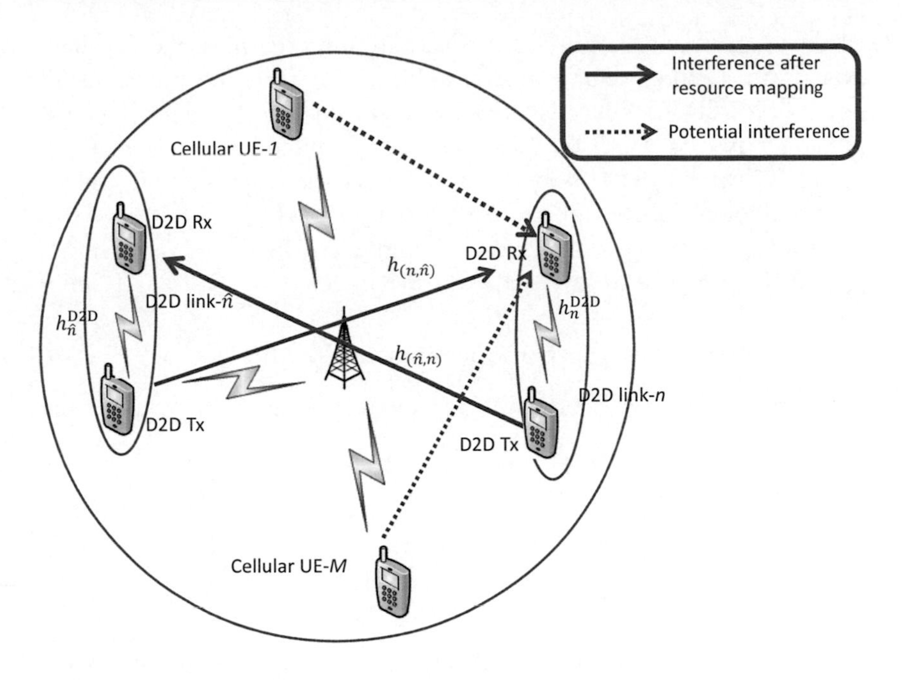

Figure 3.7: Interference scenario: D2D communication in reuse mode [JKK+14]

This point is shown in Figure 3.7 where one D2D link reuses the RB of either a cellular uplink or a D2D link. The SINRs where a D2D link and a cellular uplink reuse the same resource are already shown in Equations (3.3) and (3.4), while the SINRs that two D2D pairs reuse the same resource can be computed as

$$SINR_{(n,\hat{n})}^{\text{D2D}} = \frac{h_{n}^{\text{D2D}} P_{\text{tx}}^{\text{D2D}}}{h_{(n,\hat{n})} P_{\text{tx}}^{\text{D2D}} + \sigma_{\text{n}}^{2}}, \tag{3.23}$$

and

$$SINR_{(\hat{n},n)}^{\text{D2D}} = \frac{h_{\hat{n}}^{\text{D2D}} P_{\text{tx}}^{\text{D2D}}}{h_{(\hat{n},n)} P_{\text{tx}}^{\text{D2D}} + \sigma_{\text{n}}^{2}}. \tag{3.24}$$

These two equations show the corresponding SINR values for the nth D2D link and the $\hat{n}^{\text{th}}$ D2D link if they share the same resource simultaneously. Correspondingly, $h_{(n,\hat{n})}$ and $h_{(\hat{n},n)}$ denote the channel gains from the $\hat{n}^{\text{th}}$ D2D Tx to n^{th} D2D Rx and vice versa. As mentioned before, a

channel gain value comprises the effects from the pathloss model, shadowing factor and antenna gain. In addition, $P_{\mathrm{tx}}^{\mathrm{D2D}}$ denotes the transmit power of a D2D Tx, which is a fixed value in this section. σ_{n}^2 is the noise power at the Rx. With the SINR values computed in Equations (3.3), (3.4), (3.23), and (3.24), the achievable capacities by assigning one RB to two different links can be further derived [JF12].
With a total awareness of all channel gains, the BS can perform an improved RRM algorithm to decide which RBs are assigned to each link.

3.3.2 Construction of the optimization problem

In a cell with totally N active links, a capacity matrix $\boldsymbol{C}$ can be constructed

$$\boldsymbol{C} = \begin{pmatrix} c(1,1) & \cdots & \cdots & \cdots & c(1,N) \\ \vdots & \ddots & & & \vdots \\ \vdots & & c(m,n) & & \vdots \\ \vdots & & & \ddots & \vdots \\ c(N,1) & \cdots & \cdots & \cdots & c(N,N) \end{pmatrix}, \tag{3.25}$$

where each element $\{c(m,n), m \neq n\}$ in the matrix $\boldsymbol{C}$ represents the achievable link capacity per RB for the mth link, if it reuses the same RB with the nth link. The elements on the main diagonal of matrix $\boldsymbol{C}$, i.e. $m = n$, have a value of 0, i.e. $c(m,m) = 0$ because a link should not reuse its own resource. Besides, since it is assumed that one RB cannot be used by two cellular uplinks simultaneously, this constraint is intentionally manipulated by setting $c(m,n) = 0$, if both the mth and the nth links are cellular links. Please note that the matrix $\boldsymbol{C}$ in Equation (3.25) is different from the matrix $\boldsymbol{C}^{\mathrm{CD}}$ in Equation (3.16), since it is now assumed that two different D2D links can use the same resource. Thus, in case the mth communication link requires a minimum data rate of r_m to support its user-specific data rate requirement, an optimization problem can be formulated to minimize the number of necessary RBs to fulfill the requirements of all UEs, as

$$\textbf{minimize}_X \quad \frac{\sum_m \sum_n x(m,n)}{2} \tag{3.26}$$

$$\textbf{subject to} \quad \sum_n c(m,n)x(m,n) \geq r_m; \quad \forall m; \tag{3.27}$$

$$x(m,n) \in \boldsymbol{Z}^*; \tag{3.28}$$

$$x(m,n) - x(n,m) = 0; \tag{3.29}$$

$$m, n \in (1, 2, ..., N). \tag{3.30}$$

$$\tag{3.31}$$

The motivation for this optimization problem is to minimize the resource consumption so that additional resources can be used in other network slices to serve other types of services [HJS18]. Equation (3.26) shows the objective function to minimize the number of required RBs by searching for the optimal resource allocation matrix $\boldsymbol{X}$

$$\boldsymbol{X} = \begin{pmatrix} x(1,1) & \cdots & \cdots & \cdots & x(1,N) \\ \vdots & \ddots & & & \vdots \\ \vdots & & x(m,n) & & \vdots \\ \vdots & & & \ddots & \vdots \\ x(N,1) & \cdots & \cdots & \cdots & x(N,N) \end{pmatrix}, \tag{3.32}$$

where the value of each element $x(m,n)$ indicates the number of RBs reused by the mth and n^{th} links and, therefore, $x(m,n)$ has a non-negative integer value as indicated in Equation (3.28). If its value is 0, it means that these two links do not share the same resource. Since a link should not reuse its own resource, the main diagonal elements in the matrix $\boldsymbol{X}$ is equal to zero, as $x(m,m) = 0, \forall m$. In addition, Equation (3.29) shows that the matrix $\boldsymbol{X}$ is symmetrical and each resource is shared by two different links. Thus, if we sum up all the elements in the matrix X, the summation equals to two times of the number of the allocated RBs, which is defined as B_{required},

$$B_{\text{required}} = \frac{\sum_m \sum_n x(m,n)}{2}. \tag{3.33}$$

Thus, the objective function in Equation (3.26) calculates the required number of RBs and tries to minimize its value. As aforementioned, since the mth communication link requires a minimum data rate of r_m to support its user-specific data rate requirement, Equation (3.27) shows this constraint mathematically.

Since this optimization is an integer linear programming problem and NP-hard, there are $(N_{\text{cell}} \times N_{\text{D2D}} + N_{\text{D2D}} \times (N_{\text{D2D}} - 1)/2)^{B_{\text{required}}}$ possible resource allocation options, where N_{cell} and N_{D2D} represent the number of cellular links and the number of D2D links, respectively. If 10 cellular links and 20 D2D links are active inside a cell coverage, at least 15 RBs are required to support these 30 links by assigning each RB simultaneously to tow different links. In this case, there are approximately 7.34×10^{38} allocation options to check whether they fulfill the user-specific data rate requirement as stated in Equation (3.27). Thus, as one typical characteristic for an integer linear programming problem, a very high complexity is required for even identifying the feasible set. Therefore, in order to efficiently allocate radio resources, it is critical to developing a heuristic algorithm which yields a sub-optimal performance.

3.3.3 Heuristic RRM algorithm

In Algorithm 3, a heuristic algorithm based on the branch-and-cut algorithm is introduced to solve the optimization problem. For a good understanding, Figure 3.8 is used to demonstrate Algorithm 3. One input to this algorithm is the capacity matrix $\boldsymbol{C}$, which is defined in Equation (3.25). Moreover, another input for Algorithm 3 is the service requirement vector R defined as

$$R = (r_1, r_2, \ldots, r_m, \ldots, r_N)^T, \tag{3.34}$$

where r_m represents the data rate requirement of the mth link.

Input:

$$\boldsymbol{C}=\begin{pmatrix} c(1,1) & c(1,2) & c(1,3) & c(1,4)\\ c(2,1) & c(2,2) & c(2,3) & c(2,4)\\ c(3,1) & c(3,2) & c(3,3) & c(3,4)\\ c(4,1) & c(4,2) & c(4,3) & c(4,4)\end{pmatrix}=\begin{pmatrix} 0 & 7 & 1 & 2\\ 9 & 0 & 4 & 5\\ 8 & 6 & 0 & 3\\ 12 & 10 & 11 & 0\end{pmatrix}; R=\begin{pmatrix} r_1\\ r_2\\ r_3\\ r_4\end{pmatrix}=\begin{pmatrix} 1\\ 8\\ 12\\ 8\end{pmatrix};$$

Initialization: $A=\begin{pmatrix} 0\\ 0\\ 0\\ 0\end{pmatrix};$

$$\Rightarrow \boldsymbol{NormCp}=\boldsymbol{NormC}=\begin{pmatrix} 0 & 1 & 1 & 1\\ 1 & 0 & 1/2 & 5/8\\ 2/3 & 1/2 & 0 & 1/4\\ 1 & 1 & 1 & 0\end{pmatrix};$$

InfeasibleLinkIDs is empty; *Sum up each row* $\Rightarrow$ $Sum=\begin{pmatrix} 3\\ 17/8\\ 17/12\\ 3\end{pmatrix}$

m=3: Index of the row with the minimal non-zero summation value

Sum up each non-zero value NormC(3,n) in the 3rd row with the element of Norm(n,3) $\Rightarrow$

$NormC(3,1)+NormC(1,3)=5/3;$
$NormC(3,2)+NormC(3,3)=1;$
$NormC(3,4)+NormC(4,3)=5/4;$

$(m=3, n=1)$ generates the maximal sumation value.

$\Downarrow$

FeasibleLinkIDs $=(3\quad 1);$

Update vector A $\Rightarrow$ $A=\begin{pmatrix} 1\\ 0\\ 2/3\\ 0\end{pmatrix};$ *Update NormC* $\Rightarrow$ $\boldsymbol{NormC}=\begin{pmatrix} 0 & 0 & 0 & 0\\ 1 & 0 & 1/2 & 5/8\\ 1 & 1 & 0 & 3/4\\ 1 & 1 & 1 & 0\end{pmatrix}$

Sum up each row $\Rightarrow$ $Sum=\begin{pmatrix} 0\\ 17/8\\ 11/4\\ 3\end{pmatrix}$

m=2: Index of the row with the minimal non-zero summation value

Sum up each non-zero value NormC(2,n) in the 2nd row with the element of NormC(n,2) $\Rightarrow$

$NormC(2,1)+NormC(1,2)=1;$
$NormC(2,3)+NormC(3,2)=3/2;$
$NormC(2,4)+NormC(4,2)=13/8;$

$(m=2, n=4)$ generates the maximal sumation value.

$\Downarrow$

FeasibleLinkIDs $=\begin{pmatrix} 3 & 1\\ 2 & 4\end{pmatrix};$

Update vector A $\Rightarrow$ $A=\begin{pmatrix} 1\\ 5/8\\ 2/3\\ 1\end{pmatrix};$ *Update NormC* $\Rightarrow$ $\boldsymbol{NormC}=\begin{pmatrix} 0 & 0 & 0 & 0\\ 1 & 0 & 1 & 1\\ 1 & 1 & 0 & 3/4\\ 0 & 0 & 0 & 0\end{pmatrix}$ *Sum up each row* $\Rightarrow$

$Sum=\begin{pmatrix} 0\\ 3\\ 11/4\\ 0\end{pmatrix}$

m=3: Index of the row with the minimal non-zero summation value

Sum up each non-zero value NormC(3,n) in the 3rd row with the element of NormC(n,3) $\Rightarrow$

$NormC(3,1)+NormC(1,3)=1;$
$NormC(3,2)+NormC(2,3)=2;$
$NormC(3,4)+NormC(4,3)=3/4;$

$\Rightarrow$ FeasibleLinkIDs $=\begin{pmatrix} 3 & 1\\ 2 & 4\\ 3 & 2\end{pmatrix};$ *Update A* $\Rightarrow$ $A=\begin{pmatrix} 1\\ 1\\ 1\\ 1\end{pmatrix},$ *Update NormC* $\Rightarrow$ $\boldsymbol{NormC}=\begin{pmatrix} 0 & 0 & 0\\ 0 & 0 & 0\\ 0 & 0 & 0\\ 0 & 0 & 0\end{pmatrix}$ $\Rightarrow$

FeasibleLinkIDs $=\begin{pmatrix} 3 & 1\\ 2 & 4\\ 3 & 2\end{pmatrix};$ $\Rightarrow$ $\boldsymbol{X}=\begin{pmatrix} 0 & 0 & 1 & 0\\ 0 & 0 & 1 & 1\\ 1 & 1 & 0 & 0\\ 0 & 1 & 0 & 0\end{pmatrix}.$ $\Rightarrow$ Done

Figure 3.8: A simple example to illustrate Algorithm 3

Algorithm 3 Minimizing necessary RBs w.r.t. user satisfaction [JKK+14]

Input:
The capacity matrix $\boldsymbol{C}$, service requirement vector R.

Output:
IDs of links which cannot reuse any RB: $InfeasibleLinkIDs$.
IDs of links using the same RB: $FeasibleLinkIDs$.

1: initialize parameters:
set vector A=zeros($NumOfLinks$,1);
2: to obtain matrix $\boldsymbol{NormCp}$ and $\boldsymbol{NormC}$, normalize each row of the matrix $\boldsymbol{C}$ w.r.t. requirement vector R, i.e. $\boldsymbol{NormCp}(m,:) = \boldsymbol{NormC}(m,:) = \boldsymbol{C}(m,:)/r_m$. So far, $\boldsymbol{NormCp}$ and $\boldsymbol{NormC}$ are the same;
3: set elements in $\boldsymbol{NormCp}$ and $\boldsymbol{NormC}$, which are greater than 1 to 1;
4: **if** a set of rows S in $\boldsymbol{NormC}$ are all-zero row vectors **then**
5: $InfeasibleLinkIDs$=[indices of these rows in set S];
6: **end if**
7: RB indice $b = 1$;
8: **while** $\boldsymbol{NormC}$ is not an all-zero matrix **do**
9: sum up every row in matrix $\boldsymbol{NormC}$ and look for the row m that has the minimal non-zero result;
10: for every non-zero elements $NormC(m,n)$ in row m, add $NormC(m,n)$ with element $NormC(n,m)$;
11: look for the combination (m,n) which gives the maximal row sum value in Step 10, set $FeasibleLinkIDs(b,1) = m$ and $FeasibleLinkIDs(b,2) = n$, $b = b+1$;
12: increase a_m and a_n by $\{NormC(m,n)\times(1-a_m)\}$ and $\{NormC(n,m)\times(1-a_n)\}$, respectively;
13: **if** a_m or a_n is greater or equal to 1 **then**
14: set the ones greater than 1 to 1;
15: set the mth or n^{th} row in matrix $\boldsymbol{NormC}$ to all-zero vectors if a_m or a_n equals to 1, respectively;
16: **end if**
17: **if** a_m or a_n is smaller than 1 **then**
18: **if** a_m is smaller than 1 **then**
19: update the mth row of matrix $\boldsymbol{NormC}$ by dividing the mth row of $\boldsymbol{NormCp}$ with $(1 - a_m)$;
20: **end if**
21: **if** a_n is smaller than 1 **then**
22: update the nth row of matrix $\boldsymbol{NormC}$ by dividing the nth row of $\boldsymbol{NormCp}$ with $(1 - a_n)$;
23: **end if**
24: **end if**
25: set elements in $\boldsymbol{NormC}$ which are greater than 1 to 1;
26: **end while**
27: **for** each $i \in [1, b-1]$ **do**
28: set the elements $x(FeasibleLinkIDs(i,1), FeasibleLinkIDs(i,2)) = 1$ and $x(FeasibleLinkIDs(i,2), FeasibleLinkIDs(i,1)) = 1$ in the resource allocation matrix $\boldsymbol{X}$;
29: **end for**

In Algorithm 3, the total number of links is indicated in $NumOfLinks$ and $\boldsymbol{NormC}(m,n)$ shows the case when the mth and n^{th} links share the same RB. In addition, the vector A representing the satisfaction degree of the different links has a dimension of $(NumOfLinks \times 1)$. a_m is the mth element in vector A and defined as

$$a_m = \min\{1, \frac{\text{capacity provided by the allocated RBs for the } m^{\text{th}} \text{ link}}{\text{data rate requirement of the } m^{\text{th}} \text{ link}}\}. \tag{3.35}$$

At the beginning of the algorithm, no RBs have been assigned to any link and, thus, every element in vector A is initialized to zero. At the end of the algorithm, two output, i.e. $InfeasibleLinkIDs$ and $FeasibleLinkIDs$, will be generated, and their details are explained in the following:

- A link categorized in $InfeasibleLinkIDs$ is not allowed to share RB with another link, since its SINR value is too low to support its data transmission, no matter which other link reuses the same RB. Thus, dedicated resources should be assigned to the links in $InfeasibleLinkIDs$, and the reuse of these resources is not allowed.
- The other links shown in $FeasibleLinkIDs$ demonstrate the set of links which can share RBs with others. $FeasibleLinkIDs$ has a dimension of $B_{\text{required}} \times 2$, where each row represents which two links share the same RB, and B_{required} indicates the number of necessary RBs that will be reused.

In addition, the $\boldsymbol{NormC}(m,:)$ and $\boldsymbol{NormCp}(m,:)$ in Step 2 represent the mth row of matrix $\boldsymbol{NormC}$ and $\boldsymbol{NormCp}$, correspondingly. And they are computed by dividing the mth row of matrix $\boldsymbol{C}$ by r_m, as

$$\boldsymbol{NormC}(m,:) = \frac{\boldsymbol{C}(m,:)}{r_m}. \tag{3.36}$$

As it can be seen from Algorithm 3, Step 1 to Step 7 show the initialization and search for the links which need dedicated resources. If the mth row in the matrix $\boldsymbol{NormC}$ is an all-zero vector, the mth link can not share a resource with another link, since it achieves a capacity of 0. The elements $NormC(m,n)$ and $NormCp(m,n)$ in the matrices $\boldsymbol{NormC}$ and $\boldsymbol{NormCp}$ show that how satisfied the mth link will be, if the mth link and the nth link are assigned with a RB. At the beginning of the algorithm, these two matrices are the same. Alone with the running the algorithm, $\boldsymbol{NormC}$ will be updated, while $\boldsymbol{NormCp}$ remains. Each iteration from Step 8 to Step 26 decides which two links share the bth RB. In Step 9, the algorithm searches for the mth link, whose data rate requirement is the most difficult to be fulfilled. Following that, the algorithm checks all the combination options for the mth link, and searches for the combination that offers the maximal normalized capacity in Steps 10 and 11. In addition, once the combination is decided, the resource will be allocated to the corresponding two links in Step 11. Afterwards, the values in the vector A and in the matrix $\boldsymbol{NormC}$ will be updated. For instance, after the mth link and the nth link have been assigned with a RB, their capacity requirement will be decreased since the assigned RB provides certain capacities to these links already. Therefore, the values of the elements a_m and a_n in the satisfaction vector A should be increased as in Step 12. If the updated value of either a_m or a_n is larger than 1, it is set to 1 in Step 14 and it means that the data rate requirement of the corresponding user is fulfilled. Accordingly, the corresponding row in the matrix $\boldsymbol{NormC}$ should be set to an all-zero vector in Step 15. In the other case where the value of a_m or a_n is smaller than 1, the corresponding row in the matrix $\boldsymbol{NormC}$ will be updated as shown in Steps 17 to 24. Please note that once an RB has been assigned by using the iteration from Step 8 to Step 26, the resource allocation for this RB is considered as solid. Thus, the proposed sub-optimal algorithm requires a modest complexity that has a linear relationship with the number of necessary RBs, recalling the big amount of options in this NP-hard problem. Before the algorithm ends, Steps 27 to 29 are used to generate the resource allocation matrix $\boldsymbol{X}$.

3.3.4 Extension of the RRM algorithm: taking account of the UE priority information

So far, the author has introduced the optimization problem to minimize the number of required RBs. However, in reality it is also possible that the number of available RBs is not sufficient to fulfill the requirements of all UEs. Meanwhile, certain users in the network might be categorized as "golden" or "premium" users. Due to their service level agreements (SLAs) with network operators, these UEs expect to be served with better service quality compared to the standard UEs. In addition, some services may require higher priority, e.g. real-time services posing a strict requirement in terms of latency. Thus, a smart RRM algorithm should be capable of allocating the radio resource with the following characters:

- The links with a high priority should be served in a better manner, in order to achieve a better QoE.
- The links with low priority should not be thoroughly neglected but still served in a fair manner.

In order to meet the above two characters, Algorithm 3 is extended and modified with minor changes. One of the modifications is to add another input which is the priority vector PR defined as

$$PR = (pr_1, pr_2, \ldots, pr_m, \ldots, pr_N), \tag{3.37}$$

where pr_m states the priority level of the mth link. In this work, it is assumed that a higher value represents a higher priority level and all values should be an integer number larger or equal to 1. Moreover, four minor changes are necessary to adjust Algorithm 3, and they are listed below.

1. In Step 2, a new matrix $\boldsymbol{NormCprior}$ is introduced by normalizing each row of matrix $\boldsymbol{NormCp}$, i.e. $\boldsymbol{NormCp}(m,:)$, with the priority level of the mth link, i.e. pr_m, and then setting its elements greater than 1 to 1.
2. The operations in Step 9 and Step 10 should be now on the matrix $\boldsymbol{NormCprior}$ instead of on $\boldsymbol{NormC}$.
3. After Step 25, one more step should be added. In this step, the matrix $\boldsymbol{NormCprior}$ should be updated by normalizing each row of the matrix $\boldsymbol{NormC}$ with the priority level of the corresponding link.
4. The conditional statement "$\boldsymbol{NormC}$ is not an all-zero matrix" in Step 8 should be changed to that "$\boldsymbol{NormC}$ is not an all-zero matrix and b is not larger than the number of available RBs".

In the extended algorithm, the priority of different UEs or their service patterns are reflected in Steps 9 and 10. With these modifications, the priority of each link is taken as a weight factor to adjust its probability to compete for an RB.

Table 3.2: Simulation parameters [JKK+14, IR08a]

Case	Urban macro
Carrier frequency	2 GHz
Noise figure at BS	5 dB
Noise figure at D2D Rx	7 dB
Power level of thermal noise	-174 dBm/Hz
Cable loss	3 dB
Bandwidth per RB	180 kHz
Transmitting power of user terminal	24 dBm
Antenna patterns for all devices	omnidirectional
SINR estimation	perfect
Maximal distance between one D2D pair	120 m

Table 3.3: Context information [JKK+14]

	Number of links	Priority level	Minimal data rate
C-Voice	5	3	64 kbps
C-Web	4	2	25 kbps
C-FTP	1	1	200 kbps
D-Voice	10	2	64 kbps
D-Web	5	1	25 kbps
D-FTP	5	1	200 kbps

3.3.5 Evaluation and numerical results

The system-level simulator implemented in Section 3.2.4 is extended to generate the performance of the proposed technology. The corresponding simulation parameters are given in Table 3.2. Furthermore, the number of active links, the priority level and data rate requirement for each service type are proposed by the author and provided in Table 3.3. The parameters in Table 3.2 and Table 3.3 are used to configure the simulator and generate the results in Figures 3.9 and 3.10. In Table 3.3, the term "C-Voice" represents a voice service for a cellular link and D-Web a Web service for a D2D link, so do the other abbreviations. In addition, it is assumed that this context information is available at the BS.

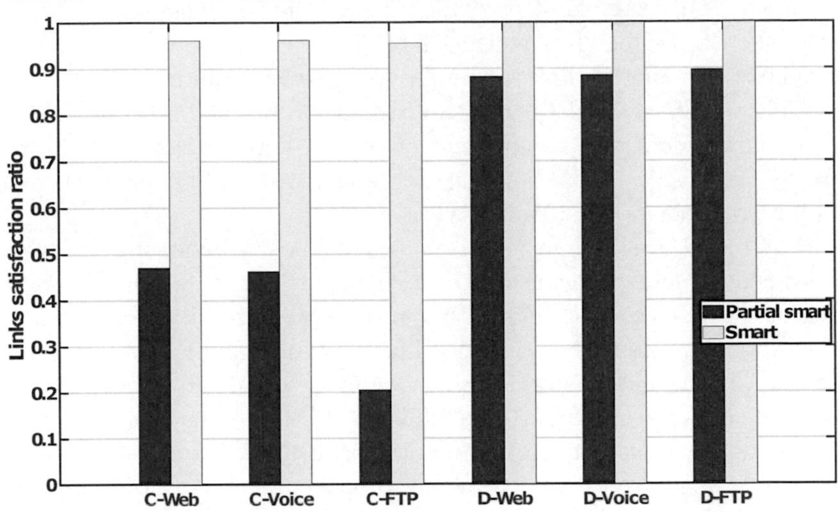

Figure 3.9: Performance comparison w.r.t. link satisfaction [JKK+14]

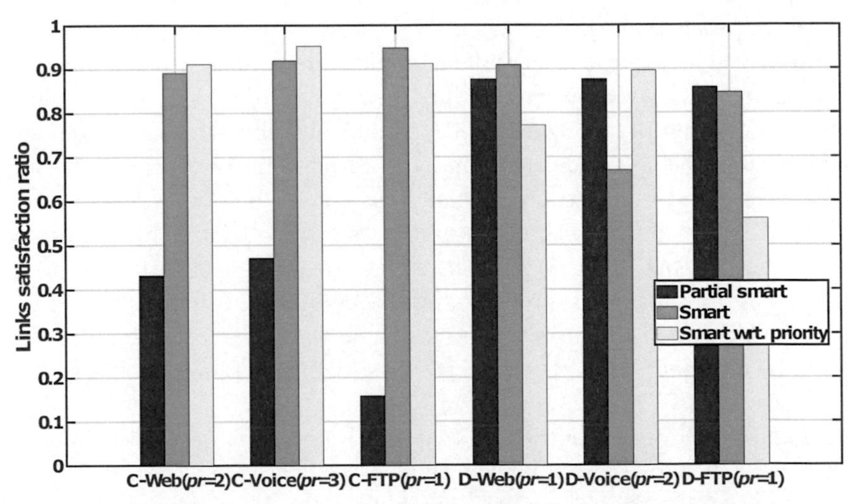

Figure 3.10: Performance comparison w.r.t. link priority [JKK+14]

As a benchmark scheme, a scheme termed as "partial smart" algorithm by the author is used for comparison. In this scheme, the data rate requirement of a service plays a critical role for the amount of allocated resources. For example, if two links request the transmission resources from the BS, i.e. one link is using FTP service requiring a data rate of 200 kbps and another link is using a Web service requiring a data rate of 25 kbps. Since the data rate requirement of the FTP service is eight times of the data rate requirement of the Web service, the probability for the link using the FTP service to be assigned an RB is eight times of the probability of the Web service.

In Figure 3.9, the system performance comparison between the developed context-aware scheme stated in Algorithm 3 and the partial smart scheme w.r.t. the link satisfaction ratio is shown. Please note that the proposed context-aware scheme illustrated in Algorithm 3 is labeled "Smart" by the author in this figure. A link is considered as being satisfied if its data rate requirement is fulfilled. Otherwise, the link is counted as not satisfied. In this scheme, no user priority level is considered and all links will be treated without any discrimination. In addition, the radio resource allocation in the partial smart scheme is performed independently of the CSI. Thus, the mutual interference is out of network control and can severely deteriorate the link quality. From this figure, it can be seen that the proposed context-aware resource allocation algorithm outperforms the partial smart algorithm. Especially, significant gains can be achieved by the cellular UEs, and this demonstrates the efficiency of the proposed algorithm to dynamically allocate radio resources by utilizing the CSI and other context information. Furthermore, it can also be noticed that certain cellular links cannot be always satisfied, if their RBs are shared with other links. As stated before, this is due to the fact that some cellular links have already low SINRs with dedicated resources, and if the resources are reused by other D2D links, the interference from the D2D links will be so strong that these cellular links can even be dropped. Thus, in this case, these cellular uplinks should go for dedicated RBs.

In Figure 3.10, together with the extended algorithm developed in Section 3.3.4 where priority information is considered, in total three algorithms are inspected with 17 available RBs. Please note that the extended algorithm developed in Section 3.3.4 is labeled "Smart wrt. priority" by the author in this Figure. The priority levels of different service types are indicated in this figure by using symbol pr and these values are listed in Table 3.3. As mentioned in Section 3.3.4, the priority level of each link is used as a weight factor by the extended algorithm to adjust its probability to compete for an RB. As shown in this figure, the links with high priorities are assigned with more RBs and their satisfaction ratio is increased by the extended algorithm compared with the other schemes. Meanwhile, as a compromise, low priority links are sacrificed but an acceptable satisfaction ratio can still be ensured.

3.4 Radio link enabler for context-aware D2D in the reuse mode

3.4.1 Challenges and the proposed solution

In the previous section, the feasibility of assigning a resource to both the cellular and D2D links, i.e. defined in Equation (3.6), has been inspected where the SINR values of the different links are taken into account, i.e. shown in Equations (3.3) and (3.4). In order to derive these SINR values, CSI of all the cellular and D2D links is required to be available at the BS. Considering a single-cell scenario where N cellular UEs and M D2D pairs exist, a total awareness of CSI implies the availability of the following information at the BS:

1. N cellular link channel gain values, i.e. h_n^{cell} in Equation (3.4), each for one uplink between a UE and the BS.
2. M D2D link channel gain values, i.e. h_m^{D2D} in Equation (3.3), each for one link between the D2D Tx and its Rx.
3. M channel gain values to characterize interference power levels for cellular uplinks, i.e. $h_{(\text{BS},m)}$ in Equation (3.4), each for one link between a D2D Tx and the BS.
4. $N \times M$ channel gain values to characterize interference power levels for D2D links, i.e. $h_{(m,n)}$ in Equation (3.3), each for one link between one uplink cellular UE and one D2D Rx.

The first two items, i.e. h_n^{cell} and h_m^{D2D}, can be obtained by measuring the DMRS at the time when a radio link is established. As aforementioned, in the sidelink transmission mode 1, a D2D Tx needs to send a scheduling request (SR) message to the BS to request the transmission resource. Along with this step, the channel gain value between the D2D Tx and the BS, i.e. $h_{(\mathrm{BS},m)}$, can be estimated by using the reference signals embedded in the scheduling request message. However, the last item, i.e. $h_{(m,n)}$, characterizing the interference power levels for D2D links can only be collected in a cumbersome manner and a large signaling overhead is foreseen to send this information to the BS.

Thus, in order to reduce the signaling overhead and achieve an efficient signaling, the UEs' location information can be exploited to estimate the interference power levels for D2D links. In this sense, it is not required to present the channel gain values of the interference links, i.e. the last item listed before, at the BS. And the interference from a cellular link to a D2D Rx is considered to be under a good control if the following equation is fulfilled:

$$\frac{d_{(m,n)}}{d_m} \geq \gamma_{(SINR^{\mathrm{D2D}}_{\mathrm{target}}, d_m, \cdots)}, \tag{3.38}$$

where d_m is the distance between the two ends of the mth D2D pair, and $d_{(m,n)}$ is the distance from the nth cellular UE to the Rx of the mth D2D pair. Besides, $\gamma_{(SINR^{\mathrm{D2D}}_{\mathrm{target}}, d_m, \cdots)}$ is a threshold value which can be configured by the network operator as a function of $SINR^{\mathrm{D2D}}_{\mathrm{target}}$, d_m, and other relevant context information. Please note that the network operator can configure this threshold value based on its own implementation strategy. For example, if the network operator wants to obtain a low interference power for the D2D link, it needs to use a high threshold value. In this thesis, the value of $\gamma_{(SINR^{\mathrm{D2D}}_{\mathrm{target}}, d_m, \cdots)}$ is set to 1 for derive the simulation performance in Section. 3.4.4. With this alternative approach for channel estimation, the feasibility function in Equation (3.6) can be replaced by a distance-based feasibility function and re-written as

$$\tilde{f}(m,n) = \begin{cases} 1, & \text{if Equations (3.2) and (3.38) are fulfilled;} \\ 0 & \text{else.} \end{cases} \tag{3.39}$$

Previously, the development of RRM algorithms has been performed to control mutual interference. However, without the support of an efficient signaling scheme in the CP, it is difficult to apply an RRM algorithm in a real system. Thus, the signaling schemes are designed in the following to support the proposed context-aware RRM algorithms with a good efficiency.

3.4.2 Signaling scheme in single-cell

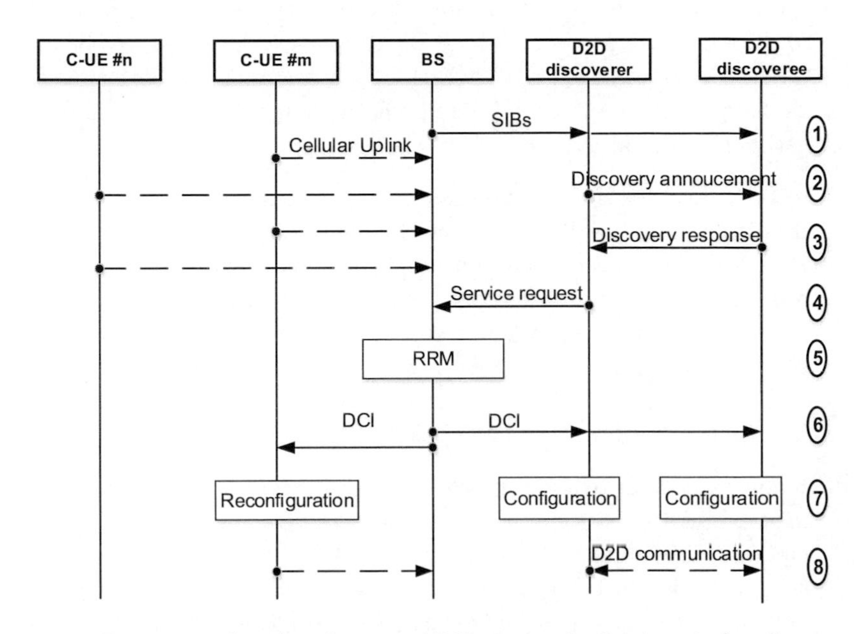

Figure 3.11: Signaling diagram of D2D communication in a single-cell

A signaling scheme is demonstrated in Figure 3.11 where two D2D ends, i.e. D2D discoverer and D2D discoveree, are served by the same BS. Please note that the D2D service authorization process is not considered since it is not the main focus of this thesis. In this signaling diagram, several cellular users are already served with dedicated resources before two nearby UEs start their direct D2D communication. In order to provide more details, the eight steps involved in the signaling diagram are illustrated with more details in the following:

1. UEs authorized to exploit D2D service receive the sidelink system information blocks (S-SIBs) broadcasted by the BS. In the S-SIBs,

configuration information for D2D discovery is carried, e.g. resource pools used for D2D discovery [3GP18f].

2. A D2D discoverer UE sends a D2D discovery message to the potential D2D discoveree UEs by exploiting either ProSe direct discovery mode A or B which are standardized in 3GPP [3GP17d]. In the mode A, D2D discoverer transmits a discovery message announcing "I AM HERE". While in mode B, a discovery message "WHO IS THERE?" or "ARE YOU THERE?" is sent. In the discovery message, application information and reference signals for D2D channel estimation are also conveyed.

3. A D2D discoveree receiving the discovery message checks the quality of the radio link by measuring the power of the reference signals. Based on the measurement outcome, the D2D discoveree announces a discovery response message to indicate its availability. If the discoveree transmits a positive response message to the discoverer, context information can be embedded in this message, i.e. position and velocity of the D2D discoveree, and its measured channel gain information for both the D2D link and the cellular link.

4. Upon reception of a positive response message, a service request message will be sent by the D2D discoverer to the serving BS, including certain context information, such as channel gain information of both the D2D and cellular links, QoS requirements of the D2D link, user positions and velocities of both the D2D ends. Please note that the D2D channel measurement is performed in Step 2, while the channel measurement of the cellular link is carried out by measuring the cell-specific reference signals (CSRS) broadcasted by the BS.

5. The BS performs the context-aware RRM algorithm, taking account of the collected context information.

6. If the RRM algorithm decides to allocate the uplink resource of a cellular UE, i.e. C-UE #m in Figure 3.11, to the corresponding D2D link, the BS sends a resource configuration message to the D2D discoverer in the downlink control information (DCI). Meanwhile, the BS may also reconfigure the cellular link whose resource

is reused in order to mitigate the impact of D2D communication in the reuse-mode.

7. After receiving the resource configuration message, the D2D pair configures its transmission with the allocated resource.

8. D2D communication starts.

Please note, in a bi-directional D2D communication case, different resources can be configured for the data transmissions in the different directions. Although the D2D discovery procedure in Steps 2 and 3 does not require both the D2D ends to enter the RRC_CONNECTED state, the D2D discoverer needs to enter the RRC_CONNECTED state to send the service request message in Step 4. The RRC_CONNECTED state is a state where a UE has already established a CP connection with the network. Corresponds to the RRC_CONNECTED state, the RRC_IDLE state is a state without an RRC connection to the network, e.g. when a UE is just powered on. Afterwards, no matter whether the D2D transmission is uni-directional or bi-directional, both the D2D Tx and Rx should listen to the DCI to receive the D2D configuration information. Thus, compared with the DCI used in the legacy LTE-A network, the DCI for D2D communication needs to carry the identity information of one D2D pair instead of one single UE.
Moreover, in case if one D2D end is out of the network coverage, the introduced procedure can still apply. However, only the D2D user under the cellular coverage can send the service request message to its serving BS. After receiving the DCI, this information will be forwarded to the other UE which is out of the cellular network coverage.

3.4.3 Signaling scheme in multi-cells

In Figure 3.12, the signaling diagram to enable D2D communication in a multi-cell scenario is introduced where the two D2D ends are served by two different BSs. The nine steps involved in this signaling diagram are detailed in the following:

1. UEs authorized to exploit D2D service receive the S-SIBs broadcasted by the BS. In 3GPP, the currently standardized S-SIBs provide resource information of not only the transmission resource

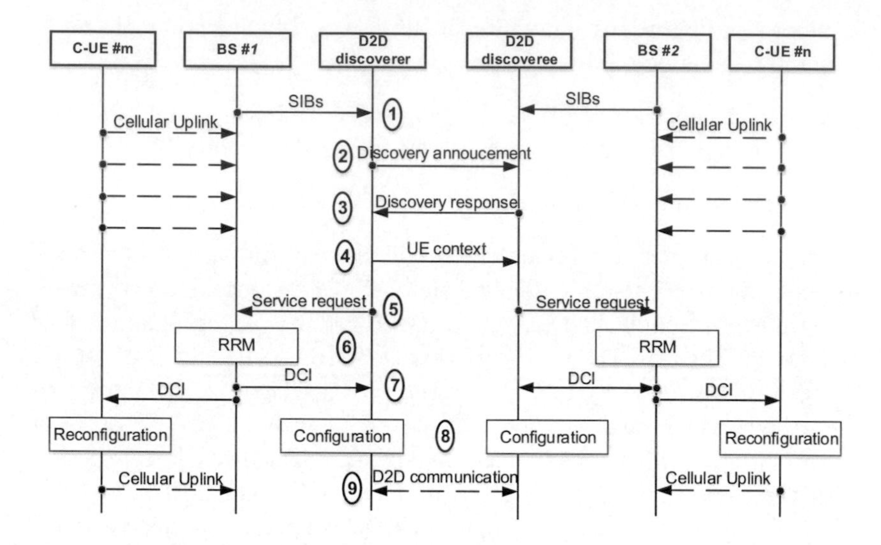

Figure 3.12: Signaling diagram of D2D communication in multi-cells

pools for the D2D discovery procedure but also the reception resource pools. Please note that the reception resource pools are normally larger than the transmission resource pools [3GP18f] so that the reception of a message from a D2D Tx served by a neighboring cell can be carried out.

2. One D2D discoverer and one D2D discoveree are paired by the ProSe direct discovery mode A or discovery mode B. Reference signals for channel estimation are contained in the discovery message. Please note that the D2D discovery message is sent over a transmission resource which is indicated by the S-SIBs received from Step 1.

3. As mentioned in Step 1, the S-SIBs broadcasted by the BS #2 contains information regarding the resource pools used for D2D discovery procedure in BS #1, therefore, the D2D discoveree can monitor over the resource where the discovery message was sent in Step 2. By estimating the D2D radio link quality, the D2D discoveree can decide whether the discovery request should be accepted or not. And a response will be sent back to the D2D discoverer, together with the context information of the D2D discoveree, e.g. CSI for

both the cellular and D2D links.

4. When an acknowledgment message is successfully received by the D2D discoverer, its context information is transmitted towards the discoveree UE.

5. Both discoverer and discoveree UEs send a service request message with the context information to their serving BSs.

6. The BSs perform the context-aware RRM algorithm, taking account of the received context information.

7. For each BS, if its associated D2D UE is allowed to transmit by reusing the resource of a cellular uplink transmission, i.e. C-UE #m and C-UE #n in Figure 3.12, a configuration message will be sent to the corresponding D2D UE. The message conveys the information about which resource is assigned to transmit and receive D2D data. Meanwhile, the cellular uplink using the same resource may also be reconfigured.

8. After receiving the configuration information from their serving BSs, two D2D ends will configure their transmissions correspondingly.

9. The bi-directional D2D communication starts.

3.4.4 Simulation assumptions and numerical results

The Madrid grid model introduced in Section 2.3 is used to generate the system performance in a dense urban scenario. As shown in Figure 2.1, a heterogeneous network deployment with both macro and micro cells is assumed. Bandwidths of 10 MHz and 40 MHz are assigned to the cellular uplinks in the macro and micro cells, correspondingly. The macro BS operates on 800 MHz while the micro BSs have a carrier frequency of 2.6 GHz. Moreover, users are distributed outdoor on streets with a density of 1000 users per $(\mathrm{km})^2$ and they are randomly selected for generating either cellular or D2D traffic. In order to compensate the propagation loss, an open loop power control is applied for both the cellular and D2D links. In this open loop power control scheme, interference is not considered and, therefore, each Tx will configure its transmit power, i.e. P_{tx}, to achieve a pre-configured SNR value at the Rx, as:

$$P_{\text{tx}} = \min\{P_{\text{max}}, P_{\text{rx}}^{\text{target}} + PL,\} \tag{3.40}$$

where P_{max} is the maximal transmit power, i.e. 24 dBm in this section, and $P_{\text{rx}}^{\text{target}}$ denotes the target reception power of the desired signal derived from the pre-configured SNR value. Last but not least, PL is the estimated pathloss value.

In order to replace $SINR_{\text{target}}^{\text{cell}}$ and $\gamma_{(SINR_{\text{target}}^{\text{D2D}}, d_m, \cdots)}$ in Equations (3.2) and (3.38), respectively, a simple method is used to set the decision thresholds for both the cellular and D2D links as

$$SINR_{(m,n)}^{\text{cell}} \geq SINR_n - \gamma_{\text{cell}}, \tag{3.41}$$

$$\frac{d_{(m,n)}}{d_m} \geq 1. \tag{3.42}$$

In this equation, $SINR_n$ represents the SINR value experienced by the n-th cellular link when its resource is not reused by any D2D link. γ_{cell} represents a deterioration offset allowed by the RRM algorithm, and it is set to 2 dB in this work.

Figure 3.13 shows the system performance of different D2D communication technologies in the case where only macro cells are deployed. Thus, the micro cells in Figure 2.1 are temporarily turned off. In addition, as the open loop power control scheme takes the pathloss into account without considering interference, the target SNR values for open loop power control are randomly generated in the intervals [10dB, 15dB] for each cellular uplink and [0dB, 10dB] for each D2D link. In Figure 3.13, the legend "Cellular Capacity" represents the capacity of all cellular links served by the macro BS, and "D2D Capacity" represents the capacity of all D2D links which operate in the reuse mode. Moreover, the legend "Overall Capacity" is the summation of the cellular capacity and the D2D capacity. Technically, three different resource allocation schemes have been implemented. In the first approach labeled as "Maximization of Overall D2D Links", the RRM algorithm tries to maximize the number of D2D links by using Algorithm 1. The feasibility function defined in Equation (3.39) will be used to construct the feasibility matrix given

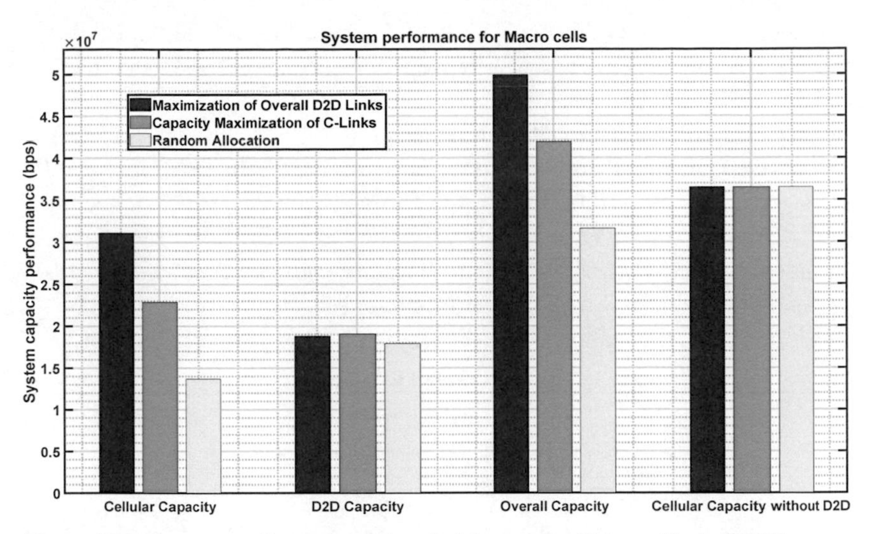

Figure 3.13: System performance with only Macro cells (Scheme No.1 of D2D power control) [JWKS17]

in Equation (3.5) which needs to be fed to the input of Algorithm 1. In this scheme, the reuse of a cellular resource is determined by the radio conditions of both the cellular and D2D links. Thus not all resources will be reused by two different links. In another scheme labeled as "Capacity Maximization of C-Links", the capacities of cellular links are maximized by using Algorithm 2. Moreover, a random allocation of cellular resources to D2D links is also investigated and denoted as the "Random Allocation" scheme in Figure 3.13. Please note that all cellular uplink resources are assigned to the D2D links in the schemes of "Capacity Maximization of C-Links" and "Random Allocation". Last but not least, as a baseline, the capacity of the LTE network is also provided where D2D communication is not allowed with the notation "Cellular Capacity without D2D". It can be seen from Figure 3.13, due to the mutual interference coming from D2D links in the reuse mode that the cellular uplink capacity in the "Maximization of Overall D2D Links" scheme has approximately decreased by 16% compared with the case where D2D communication is not allowed. Nonetheless, the overall capacity has a performance gain of approximately 35% thanks to the capacity introduced by the D2D communication. In comparison, the other

two scheme“Capacity Maximization of C-Links” and “Random Allocation” have lower gains w.r.t. the overall capacity. However, since all cellular resources are reused in the scheme “Capacity Maximization of C-Links”, its capacity of D2D links is higher than that of the scheme “Maximization of Overall D2D Links”. In turn, the additional D2D links introduce more interference to cellular links and, thus, the cellular links experience a more severe performance deterioration. Moreover, it can also be seen that the random allocation scheme provides the least performance gain due to a lack of intelligence.
In Figure 3.14, the same system setting is applied as before, except that

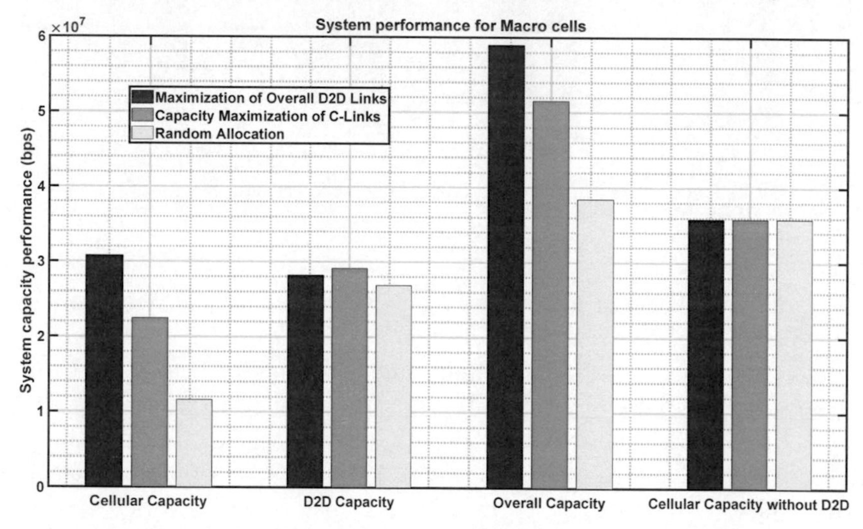

Figure 3.14: System performance with only Macro cells (Scheme No.2 of D2D power control) [JWKS17]

the target SNR value for the open loop power control has now been increased to an interval of [7dB, 12dB] for D2D links. It can be seen that the cellular links in the proposed scheme “Maximization of Overall D2D links” experience a capacity deterioration of 15%, which is approximately the same as in the previous case shown in Figure 3.13. The reason is that the same interference control scheme for cellular links is applied in the RRM algorithm. However, the increased transmit power of D2D links contributes to a higher D2D capacity compared with the previous case and, therefore, the overall capacity has an increase of 64%

compared with the legacy 4G network where D2D communication is not allowed.

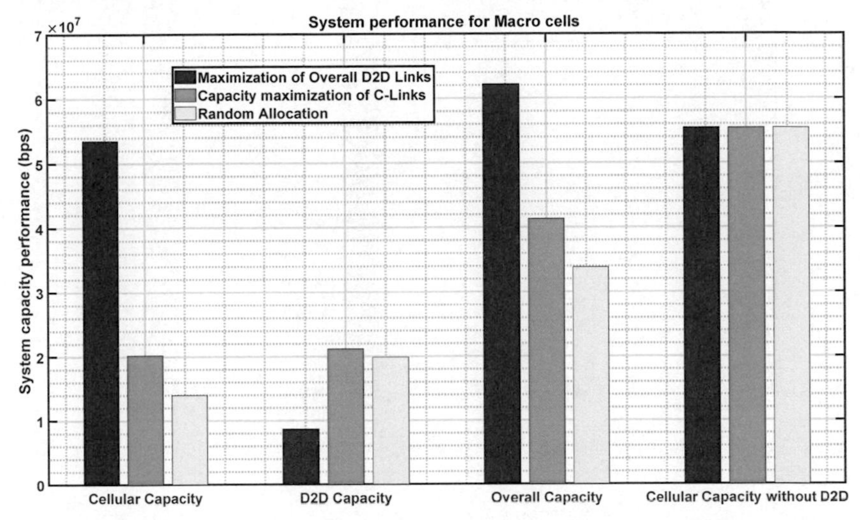

Figure 3.15: System performance for macro cells in heterogeneous network [JWKS17]

From now on, the micro cells in Figure 2.1 are turned on, and the author inspects the performance of the proposed network-controlled direct D2D communication in a heterogeneous network. The target SNR value of each D2D link is randomly and uniformly distributed in [0dB, 10dB]. In Figures 3.15 and 3.16 the system performances of macro cells and micro cells are demonstrated, respectively. In the case where D2D is not allowed, the cellular capacity is now higher than in the previous case. This is due to the network densification by deploying the micro cells, and users experiencing low SINR values from a macro cell will now be served by micro cells. In other words, if users locate closely to a macro cell, they are attached to the macro cell. Otherwise, they would be served more likely by the micro cells. Thus, if a D2D user served by the macro cell is allocated to reuse the cellular uplink resource, a high interference power level will be introduced for the cellular uplink. Consequently, less D2D links are allowed to reuse the cellular uplink resource, and the overall capacity has an improvement of 13% from the scheme “Maximization of Overall D2D Links” which is less than that shown in Figure 3.13. On the other hand, the performance of the cellular uplink will be less

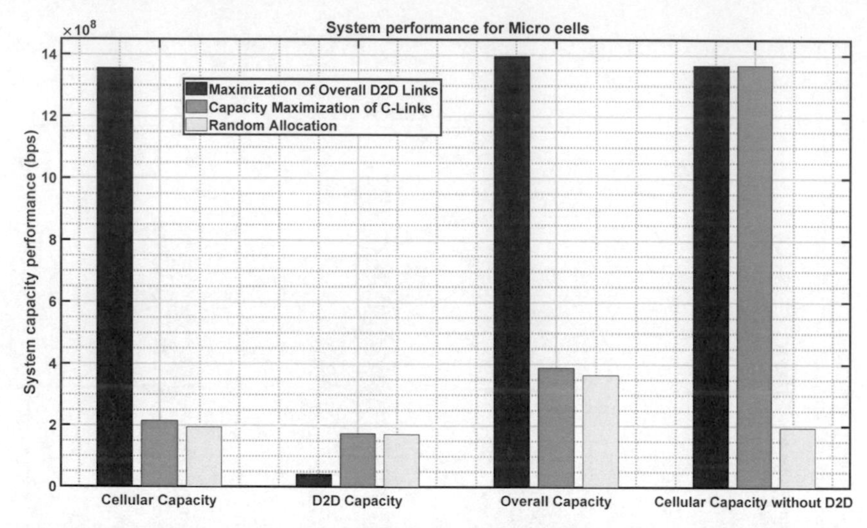

Figure 3.16: System performance for micro cells in heterogeneous network [JWKS17]

impacted since less D2D links reuse the same resource. Therefore, the capacity of cellular links will only be slightly decreased compared with the legacy 4G network. Moreover, due to the existence of micro cells, the coverage area of each cell is much smaller than the case where only macro cells are deployed. Thus, the mutual interference between cellular and D2D links can be so severe that the deterioration of radio conditions needs to be controlled more properly. This explains why the overall capacities of the other two schemes "Capacity Maximization of C-Links" and "Random Allocation" are even lower than in the case where no D2D communication is applied. In Figure 3.16 the capacity of micro cells is demonstrated. Due to a small coverage area of each micro cell, very few cellular resources can be reused by D2D links and a capacity increase of only 4% is foreseen by using the scheme "Maximization of Overall D2D Links".

3.5 Summary

In this chapter, the author has proposed a network-controlled direct D2D communication to offload traffic volume from the cellular networks to a direct communication between two user devices. As a key topic in D2D communication, the author has designed different RRM algorithms with low computational complexity. In order to improve the efficiency of resource allocation, context information has been taken into account, e.g. channel gain information, UE location, UE data rate requirement, etc. The developed RRM schemes enable an efficient resource allocation to D2D communication in the reuse mode and for efficiently achieving optimal solutions regarding different KPIs such as

- to maximize the number of feasible D2D links;
- to maximize system capacity, e.g. cellular capacity, D2D capacity, and overall capacity.

The low computational complexity of the proposed heuristic algorithms empowers a network to set up and configure D2D links with a valid response and reduces the setup delay. Numerical results have been provided where the developed resource allocation algorithms yield a better system performance in terms of the number of established D2D links, i.e. a performance gain of 68% in the considered scenario, compared with the baseline RRM scheme, and the system capacity, i.e. a performance gain of 20% in the considered scenario, compared with the baseline scheme. In addition, since spectrum is a valuable resource in wireless communication, a smart RRM algorithm has also been proposed to minimize the number of required RBs to support the service requirements of different links. Furthermore, the smart RRM algorithm is able to take additional context information into consideration, such as service requirements and user priorities and, therefore, improves the QoE. The proposed RRM algorithm can be used to determine whether a link should operate in a dedicated mode or in a reuse mode. If a D2D link operates in the reuse mode, this algorithm further decides which other link shall share the same RB. The provided numerical results have shown a better system performance by the proposed algorithm compared with a partial smart algorithm in terms of the link satisfaction ratio. To be more specific, the proposed algorithm assigns more resources to the users and services that have higher priority so that they experience better service quality. In comparison, the service quality for the users and services with lower priority is sacrificed. However, an acceptable satisfaction ratio can still be ensured for them.

From a system design perspective, efficient signaling schemes are necessary to support any context-aware RRM algorithms. Thus, signaling schemes have been designed to support the corresponding RRM algorithm in both the single-cell and multi-cell cases. Numerical results generated from a realistic scenario show that the RRM algorithms with the support from the proposed signaling schemes yield an improved system performance for macro cells, i.e. 64% in the considered scenario, compared with the 4G network in terms of the overall system capacity.

Chapter 4
Applying Network-controlled Sidelink Communication in the mMTC Use Case

4.1 Introduction

The upcoming 5G cellular system considers the mMTC as an important type of emerging services which will open a new potential market [LCL11, 3GP13c]. In 5G, mMTC normally refers to a scenario where a large number of MTDs sporadically report to an application server, e.g. to monitor environment or to track the object condition. However, since legacy cellular networks, e.g. 3G and 4G, were mainly designed to offer human-driven services, they face technical challenges to fulfill the corresponding requirements, such as low device costs and long battery standby time, e.g. ten years. A long battery standby time enables the operation on device battery without changing the device battery frequently.

In order to achieve a good efficiency and obtain a deep market penetration in applying 5G to provide mMTC services, 3GPP has carried out research to adapt and evolve the legacy cellular networks. For example, a new type of UE, i.e. the UE category 0, is introduced in [3GP16d] in order to reduce the device complexity by reducing the bandwidth and peak data rate, also using a single antenna. Moreover, the integration of the PA can be simplified by reducing the maximal transmit power, and this contributes to a further cost reduction [3GP13c]. On the other hand, a low transmit power reduces the transmission range. In addition, since some mMTC devices are deployed deep-indoor, they can experience a penetration loss up to 20 dB [3GP15a]. Due to the low transmit power and high penetration loss, the network uplink coverage is limited. Thus, in order to maintain a good uplink coverage and network availability, both narrow-band transmission and TTI bundling [3GP15a] techniques have been proposed. However, they both lead to a large transmission time and, therefore, to a huge battery energy drain at the device. Thus, it is critical to develop a mechanism with a better energy efficiency for supporting different deployment scenarios, e.g. in both rural and urban areas.

As a technical enabler for 5G, the original motivation of D2D commu-

nication was to enable a local information exchange to obtain a low latency [DRW+09, dSFM14, OMM16]. So far, D2D communication has been standardized in 3GPP to enable a direct discovery and communication procedure between two nearby devices without involving network infrastructures in the UP [3GP17d]. Moreover, in 3GPP Release 13, a UE-to-network relaying was introduced and viewed as a special type of D2D communication to extend network coverage in the public safety use case and to assist automatic driving. In this scheme, the BS controls the selection of a UE-to-network relay by setting requirements on the radio quality that a relay UE should fulfill. A remote UE that is out of the network coverage needs to measure the radio quality from different UE-to-network relay UEs and locally perform the ranking among them. The ranking procedure can be carried out by selecting the relay UE with the best radio quality to the remote UE. Although the proposed scheme is able to extend network coverage and support public safety communication, it is not optimized for the mMTC use case.
Recently, a network-controlled D2D communication has been proposed to provide an alternative to enhance the mMTC services. In [PP13, PP14], one cellphone is assumed to act as a relay to set up D2D communication for other sensors and, therefore, the packets generated by the remote sensors out of the cellular uplink coverage can be transmitted to the BS by traveling through the relay UE. Moreover, the feasibility to apply D2D communication for mMTC applications is also studied by industry in [3GP15b], where a cellphone autonomously decides to act as a relay for a group of mMTC sensors. However, in this proposal, since the D2D pairing procedure needs to be performed in a distributed manner without the assistance from the network, only a local optimization is possible and it leads to a loss in the global optimization. Moreover, the proposed approach has a limited applicability due to the fact that very few human-carried cellphones will appear deep-indoor or in rural areas, especially at night.
Due to a higher maximal transmit power of the BS compared to a user device, the transmission in downlink between a specific UE and its serving BS can have a higher SNR value than its transmission in uplink. Thus, the author inspects how to improve the network uplink coverage to serve remote mMTC sensors and meanwhile have more sensors meeting the long battery life requirement in this chapter. The sensors

deployed deep-indoor, e.g. in the basement of a building, or remotely at the cell border, e.g. far away from the serving BS in rural areas, are referred as remote sensors, since they experience a high pathloss for cellular links. Additionally, it is assumed that a sidelink communication can be performed between two nearby mMTC sensors. Thus, a remote sensor is able to set up a sidelink communication with a relay sensor. Along with the proposed technology, some insights for the further standardization work to apply sidelink communication for mMTC service will also be provided. Basically, the technology developed by the author in this chapter differs from the current work of 3GPP [3GP17d] in the following points:

- The focus is on a smart transmission mode (TM) configuration, i.e. the configuration of the relay and remote UEs, in the mMTC use case. So far in 3GPP, the corresponding technology is mainly designed for a public safety use case.
- In the proposed scheme, D2D groups are pre-selected and configured by the BS, while in the current approach in 3GPP the remote UE locally performs the ranking of the potential UE-to-network relays.
- In this work, mMTC-related context information, e.g. sensor location, battery level, and service requirement, is collected with a reasonable signaling overhead and exploited to improve network coverage and energy consumption of the sensors.

To describe the proposed scheme, this chapter is organized as follows: First, the system model for the proposed scheme is introduced in Section 4.2. Then in Section 4.3 a context-aware clustering scheme is provided to guarantee a good sidelink transmission efficiency. Moreover, in order to configure each sensor properly, a smart TM selection procedure is also stated in this section. Following that, the signaling schemes to support the proposed sidelink communication are demonstrated in Section 4.4, and numerical results generated by a system-level simulator will be shown in Section 4.5. Finally, this chapter will be summarized in Section 4.6.

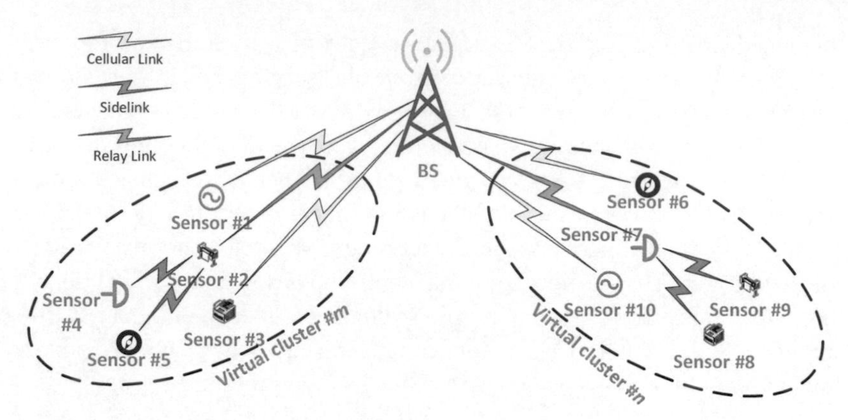

Figure 4.1: Assigning sensors to relay packets from remote sensors [JHLS17]

4.2 System model

As mentioned before, in a typical mMTC use case, many sensors need to sporadically report their status or measurements with a small payload in the uplink to the application server. In the downlink, the transmission from a BS to a sensor can be used to transmit system control information and also to page sensors for triggering an uplink report. As mentioned before, due to a higher maximal transmit power at the BSs in a cellular network, the downlink transmission usually has a better coverage than the uplink transmission and, therefore, this thesis focuses on the uplink performance.

In Figure 4.1, the proposed scheme to exploit sidelink communication to enhance mMTC services is shown. It is assumed that network has enough time-and-frequency resource, and the main problem is how to improve network coverage and energy consumption. Moreover, the signaling schemes to collect the locations of sensors are shown in Section 4.4 and, therefore, the locations of sensors can be known at the BS. As can be seen from this figure, due to a smaller signal propagation distance to the BS, sensor #2 experiences a better radio propagation channel to the BS than sensors #4 and #5. Thus, if the battery level of sensor #2 is high enough, the BS can configure it to relay the packets from other sensors, i.e. sensors #4 and #5, to the BS by applying sidelink communication. In addition, in order to obtain a good efficiency, context information, e.g. location, radio quality and battery levels of the sensors, can be exploited by the network to configure the sidelink communication. For example, a sensor in the virtual cluster $\#m$ in Figure 4.1 is not allowed having a sidelink communication with a sensor in another virtual cluster, e.g. the virtual cluster $\#n$. Thus, a good efficiency for D2D communication can be achieved by ensuring that the two sensors of a sidelink communication are geographically located near each other. As the figure shows, three TMs exist for mMTC sensors as

- cellular TM, where sensors upload their data reports to the BS directly via the cellular uplink;
- relay TM, where sensors receive reports from other sensors via sidelink communication and then transmit both the received packets and their own reports to the BS via the cellular uplink;
- sidelink TM, where sensors transmit reports to corresponding relay sensors via sidelink communication.

4.3 Device clustering and transmission mode selection

To ensure that a sidelink communication only takes place between two nearby sensors, an efficient approach is to first perform a device clustering algorithm and then only allow intra-cluster sidelink communication. Therefore, the context-aware sidelink communication scheme proposed by the author can be divided into two steps:

1. Clustering of mMTC devices;
2. selection of TM.

Once a sensor is attached to a BS or the BS needs to update the TMs of its serving sensors, the above two steps should be carried out at the BS.

4.3.1 Clustering of mMTC devices

An efficient clustering algorithm can assist a remote sensor to find a suitable relay sensor in its proximity. In order to improve the efficiency, the clustering algorithm can take useful context information into account,

such as geographical device information, traffic type, and battery life requirement. For example, the remote sensors in a rural area are normally far away from the BS. Then, the BS can analyze the geographical location of its serving sensors and configure a sensor located between the remote sensor and the BS as a relay. In another case where sensors are deployed in a dense urban scenario, sensors located deep-indoor or in basements are considered as remote sensors due to their experienced high penetration loss. In this case, the sensors located on the top floors of buildings are considered as potential relays by the BS. Thus, an efficient clustering algorithm should have the capability to dynamically adapt to the concrete situations.
In the following, the clustering algorithms will be detailed w.r.t. the environment, i.e. rural area or dense urban area. Please note that for sensors locating near the BS their propagation losses are relatively low. Thus, they can directly transmit their reports to the BS via the cellular uplink, and sidelink communication may not be applied for these UEs. This area without applying sidelink communication can be represented by a circle with radius R_{in}.

4.3.1.1 Clustering in a rural area

In a rural area, it is assumed that sensors are deployed on a 2D plane and thus a K-means clustering algorithm can be applied w.r.t. the reference angles of the sensors. In this scheme, the geographical locations of the sensors are first converted to their reference angles by the BS. After that, the reference angles will be fed to the input of the K-means clustering algorithm that carries out the following steps:

1. Initially, the BS selects the K sensors with reference angles separated from each other as far as possible; these sensors are considered as the centroids of the K clusters.

2. After that, the BS randomly picks another sensor and associates it to the cluster, whose centroid is the closest to the selected sensor w.r.t. the reference angle.

3. The BS computes the mean reference angle of the updated cluster, and selects the sensor in this cluster nearest to the mean value as the new centroid.

4. The BS iterates Step 2 and Step 3 until each sensor is associated with a cluster.

Please note that the value of K denotes the number of clusters in the K-means clustering algorithm. This is a value decided by the BS. For instance, K can be derived by dividing the cell coverage area by the area per sector. In addition, since it is assumed that the mMTC sensors are static and their positions will not change over the time, the BS can acquire the geographical information of a sensor at its initial attachment to the network, as stated in Section 4.4.

4.3.1.2 Clustering in a dense urban scenario

In an urban area, the majority of mMTC sensors will be distributed on different floors of buildings [3GP15a]. Due to this, the channel propagation condition comprises the penetration loss, and the clustering algorithm needs to be accordingly adjusted. In this part, the author introduces four different approaches for sensor clustering.

Geometrical clustering

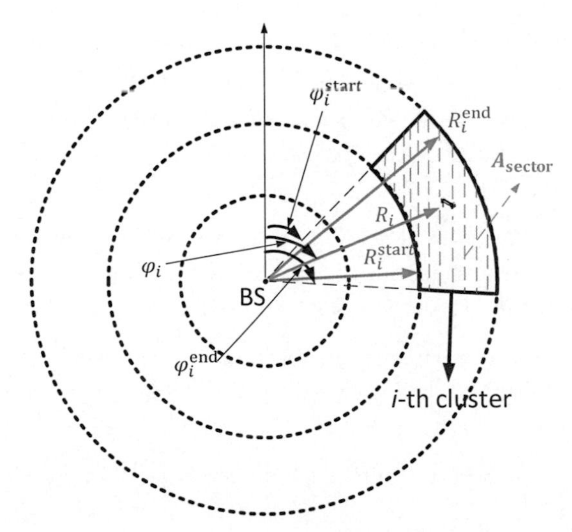

Figure 4.2: Geometrical clustering in a dense urban scenario [JLS17]

In this approach, the coverage area of the BS is sectored geographically. As shown in Figure 4.2, the area covered by the i-th cluster can be represented by the radius R_i and the reference angle φ_i as

$$R_i^{\text{start}} < R_i \leq R_i^{\text{end}}, \tag{4.1}$$

and

$$\varphi_i^{\text{start}} < \varphi_i \leq \varphi_i^{\text{end}}. \tag{4.2}$$

In Equation (4.1), R_i^{start} and R_i^{end} represent the distances from the reference point, i.e. the BS, to the inner and outer circles of the i-th cluster. In addition, φ_i^{start} and φ_i^{end} denote the reference angles between which the i-th cluster covers. The number of clusters to cover the whole coverage area is a function of A_{sector}, which is the area of one cluster. Please note that all clusters in a cell have the same size A_{sector} in the geometrical clustering scheme in this thesis. Later in Section 4.5, different values of A_{sector} are applied to yield the system performance.

K-means clustering

The K-means clustering algorithm is also applied to group sensors in a rural area. However, instead of using reference angles, the input to the clustering algorithm is the geographical location information of the sensors, which makes a difference from the K-means clustering algorithm in Section 4.3.1.1. The basic steps are listed below:

1. Initially, K devices located as far away as possible from each other in the horizontal plane are considered as the centroids of the K clusters.

2. After that, another device is randomly selected and associated to the cluster, the centroid of which has the shortest distance in the horizontal plane from the selected device.

3. The BS computes the mean coordinates of the newly formed cluster and considers the device closest to the mean coordinate as the new centroid of this cluster.

4. Iterate Steps 2 and 3 until each device is associated with a cluster.

Distance-based clustering

This scheme is similar to the K-means clustering algorithm, as detailed in the following steps:

1. K devices are randomly selected as the centroids of the K clusters.
2. Take another device randomly and associate it to the cluster which has the shortest distance from its centroid to this device.
3. Iterate Step 2 until all devices are associated with clusters.

It should be noticed that, at the beginning of this algorithm, the centroids are selected randomly from the data set and they are not updated during the operation of this algorithm. It shows the difference of this scheme compared with the K-means clustering method.

Distance-plus-CSI-based clustering
Compared with the previous distance-based clustering scheme, this approach does not only consider the location information. Moreover, the CSI of the cellular uplink between each sensor and the BS is taken into account by the BS, and the K centroids can only be selected from the devices with cellular uplink SNR values higher than a pre-defined threshold. The way to predict the uplink SNR values will be discussed later in Section 4.4.

4.3.1.3 A brief summary of the different approaches

As mentioned before, the design of an efficient clustering algorithm is highly relevant for the achievable performance of the proposed sidelink communication technology in the mMTC use case. The clustering algorithm should fulfill the following objectives:

- Each device cluster should contain the devices with different cellular radio qualities. Since only intra-cluster sidelink communication is allowed in this thesis, there will be no relay candidates if a cluster covers only the devices with bad radio conditions.

- The sidelink pathloss between two devices in one cluster should not be too large. In a cluster with a lot of sensors, a huge signaling effort is required to collect all the pathloss values among all the devices in that cluster. Therefore, as an alternative to control the sidelink pathloss, the location information of sensors can be used to ensure that a cluster only contains the devices located nearby each other.

According to the characters of the different approaches discussed before, they can be categorized as follow:

- The approaches use sensor positions. This category includes the clustering algorithm in a rural area, as introduced in Section 4.3.1.1. In addition, this category also contains the geometrical clustering approach, the K-means clustering approach, and the distance-based clustering approach shown in Section 4.3.1.2.
- The approach exploits both the position and the channel status information of a sensor, i.e. the distance-plus-CSI-based clustering approach in Section 4.3.1.2. In an urban scenario, the radio condition of a link does not only depend on the signal propagation distance. More factors should be taken into account, such as the penetration loss due to the indoor deployment of sensors. Thus, it is challenging for a BS to estimate their radio conditions by only using the positions of sensors in an urban scenario, and the reporting of the channel status information from a device is important.

The introduced clustering algorithms will be implemented the simulator and their performance will be shown and discussed in Section 4.5.

4.3.2 TM selection

After performing the clustering algorithm, the BS needs to configure each sensor with a proper TM. Again, context information can also be applied to optimize the TM configuration. For instance, a high pathloss value of a cellular link implies a large energy consumption per packet transmission via the uplink and, therefore, the long battery life requirement cannot be fulfilled. In this case, a relay sensor is needed to improve the battery usage of the remote sensor. In addition, some sensors might be even out of the cellular network coverage in the uplink due to their extreme high pathloss values, and relay sensors should be assigned to enlarge the network coverage. However, not all sensors with good propagation condition in the uplink are suitable to act as relays, since the radio quality of the sidelink is also highly relevant for energy consumption at a remote UE. Moreover, during the sidelink discovery process, a sidelink discovery message is sent by each relay sensor and, thus, the number of relay sensors should not be too large in order to keep the signaling overhead under good control. In this thesis, the TM selection for each sensor is carried out as follows:

1. Sensors with battery life lower than the requirement and sensors out of the network coverage, i.e. as given later in Equation (4.3), are considered as remote sensors and configured with the sidelink TM.

2. Sensors with good cellular radio conditions and enough battery capacities are seen as relay sensor candidates.

3. In a rural area, for sensors in each cluster a relay sensor candidate with a distance to the BS larger than a given threshold is randomly selected as the relay sensor. In an urban area, the BS randomly selects a sensor from all the relay sensor candidates as the relay in the considered cluster.

4. A configuration message is sent out by the BS to both the relay and remote sensors to trigger the sidelink discovery procedure.

5. In the sidelink discovery procedure, the pathloss of the sidelink will be estimated by each remote sensor, and a remote sensor remains in the sidelink TM, if the sidelink channel pathloss is below a threshold. Otherwise, it is configured to reside in the cellular TM.

6. Other sensors not involved in the sidelink discovery procedure are configured with the cellular TM.

Please note that the pre-condition regarding battery life stated in Step 1 can be mathematically described as:

$$t_{\mathrm{A}} + t_{\mathrm{F}}^{\mathrm{cell}} \leq BL_{\mathrm{threshold}}. \tag{4.3}$$

In this equation, $BL_{\mathrm{threshold}}$ denotes the battery life requirement, e.g. 10 years. t_{A} stands for the amount of time that the device has been served, and $t_{\mathrm{F}}^{\mathrm{cell}}$ shows the amount of time that the device can be served if the cellular uplink is used. $t_{\mathrm{F}}^{\mathrm{cell}}$ can be calculated as

$$t_{\mathrm{F}}^{\mathrm{cell}} = \frac{BC_{(i,j)}(t)}{EC_{(i,j)}}, \tag{4.4}$$

where $BC_{(i,j)}(t)$ refers to the real-time battery capacity of user-j in the cluster i at time instance t. Besides, $EC_{(i,j)}$ is the energy consumption

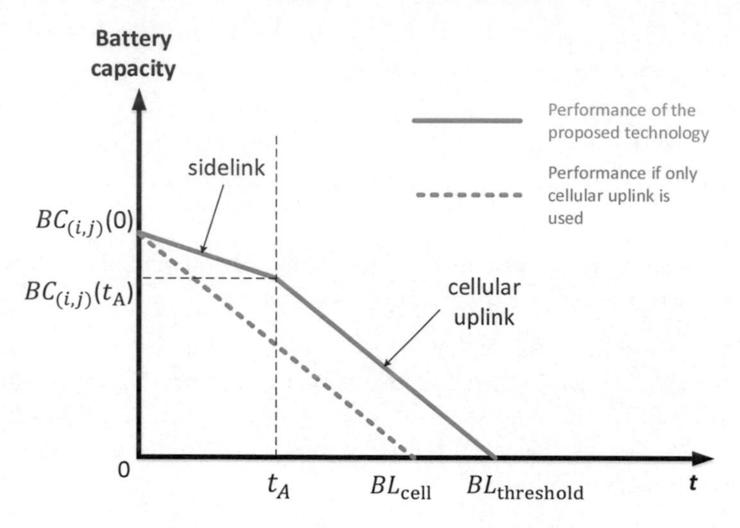

Figure 4.3: Illustration of the remote sensor battery performance

by using the cellular uplink for a unit time. For a better understanding of Equations (4.3) and (4.4), Figure 4.3 is used to illustrate the battery performance of a remote sensor. In this figure, $BC_{(i,j)}(0)$ stands for the full battery capacity at the time instance $t = 0$. The dashed blue line shows the battery performance of using cellular uplink, while the green solid line is used to show the performance of the proposed sidelink communication scheme. As shown in this figure, BL_{cell} stands for the battery life of the sensor if only the cellular uplink is applied. Since $BL_{\text{cell}} < BL_{\text{threshold}}$ and the condition in Equation (4.3) is fulfilled, the sensor cannot meet the battery life requirement by using the cellular uplink and it will be considered as a remote sensor by the BS. Therefore, at time instance $t = 0$, the network will configure the remote sensor to set up a sidelink communication with the relay sensor. With the sidelink communication, its pathloss is lower than that of the cellular uplink and, therefore, the battery of the sensor experiences a slower energy drain. After a time period t_{A}, the inequality in Equation (4.3) is not fulfilled any more, and the BS finds out that the cellular uplink can support the sensor with a battery life meeting the requirement. Thus, this sensor will not use sidelink communication from now on and it will stay in the

cellular TM as shown in Figure 4.3.
Moreover, the time difference between two successive transmission mode selection (TMS) update procedures denoted as Δt is directly related with the amount of signaling overhead used to correspondingly configure the remote and relay sensors. In order to achieve a good compromise between the configuration flexibility and the signaling load, Δt should have a value ranging from several hours to several days, depending on the traffic models of the sensors. To be more specific, the period of the TMS update procedure should be at least the same as the minimal uplink transmission period of the mMTC service. Otherwise, multiple TMS update procedures can occur between two consecutive reports of a sensor, while only the last TMS update procedure makes sense. On the other hand, a very large TMS update period reduces the network flexibility to cope with real-time condition changes. For example, since a relay sensor forwards the packets of a group of sensors, it experiences a high battery energy drain. In this sense, the BS needs to select a new relay sensor in a timely manner. Otherwise, the old relay sensor might power out, and the remote sensors will lose their connections. In addition, an infinite value is assumed for $EC_{(i,j)}$ if the user is out of network coverage in the uplink. With this assumption, sensors out of coverage also meet the condition shown in Equation (4.3) and, thus, sidelink communication is applied to serve their transmissions. Also, the pre-conditions in Step 2 can be represented as

$$\frac{BC_{(i,j)}(t)}{EC_{(i,j)}} > BL_{\text{threshold}} - t, \tag{4.5}$$

and

$$SNR^{(i,j)} \geq SNR_{\text{threshold}}. \tag{4.6}$$

Equation (4.5) shows that user-j in cluster i can meet the battery life requirement by using cellular uplink transmission. In other words, this user has enough battery capacity to serve as a relay for remote UEs in the cluster i. In Equation (4.6), $SNR^{(i,j)}$ is the estimated SNR value of the cellular uplink, and $SNR_{\text{threshold}}$ is a threshold value to check whether the channel condition of the cellular uplink is good enough. Setting a higher threshold value can force the BS to select a relay sensor with a better radio condition in the uplink and, therefore, it further contributes to a higher spectral efficiency. On the other hand, a higher value of $SNR_{\text{threshold}}$ also means less available relay candidates, and there is a high risk that no suitable relay candidate exists in a cluster. Please also notice that in Step 3 the remote sensors in a rural area are far away from the BS. Therefore, to guarantee that each relay sensor locates in the proximity of the remote sensors in the considered cluster, the relay candidates need to have a distance to the BS larger than a network-configured threshold.

4.4 Design of signaling procedures

To support the proposed context-aware sidelink communication in the mMTC use case, the required signaling procedures are designed by the author in this section. Three important procedures including the initial attachment, the update of the TM, and the uplink transmission exploiting sidelink communication are discussed in detail.

4.4.1 Initial attachment of a sensor

Once a sensor is deployed for the first time and powered on, an initialization procedure shown in Figure 4.4 is proposed by the author. Please note that the new sensor shown in Figure 4.4 is configured with sidelink TM, and sensor $\#m$ is the corresponding relay sensor. The signaling diagram is illustrated in more detail in the following:

1. The new sensor tries to receive the S-SIBs which contain the configuration information for the sidelink operation. For instance, by receiving the S-SIBs, a sensor can be aware of whether sidelink discovery and communication are supported by the network.

2. After that, the new sensor needs to report its measured CSI, location and battery level information to the BS in order to perform initial attachment to the network. Since the deployment of the new sensor is performed manually by the technicians, such information can be submitted through the equipment of technicians, though the sensor might be out of the uplink coverage. In addition, the BS can acquire other context information of a sensor, e.g. its traffic type, from an application server deployed in the core network of an operator.

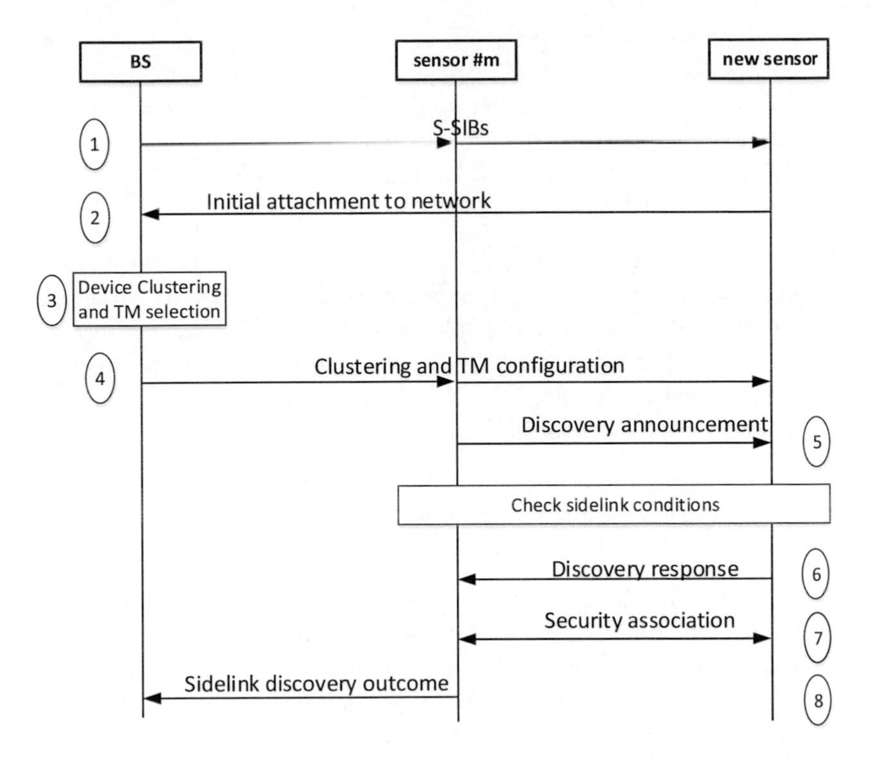

Figure 4.4: Initial attachment of a new sensor

3. Afterwards, the sidelink device clustering and the TMS will be performed by the BS, taking account of the collected context information.

4. In case the new sensor is configured by the BS with either a relay TM or a sidelink TM, the identity (ID) of its virtual cluster and other control information for sidelink discovery and communication will be transmitted from the BS to both ends of the sidelink, e.g. transmit power, IDs of the sidelink counterparts, and the resources to transmit the discovery message.

5. If the new sensor is configured to the sidelink TM while another sensor #m acts as its relay, sensor #m transmits a sidelink discovery announcement containing the ID of the new sensor. In another case,

if the new sensor is configured to the relay TM to serve a remote sensor, a sidelink discovery announcement message will be sent to that remote sensor. In this message, the IDs of sidelink counterparts and reference signals are also embedded. Please note that the time and frequency resources used to transmit the discovery message are indicated in the message received in Step 4.

6. Upon receiving the sidelink discovery announcement, the Rx will estimate whether the sidelink reference signal received power (S-RSRP) is above a threshold value or not. Based on this estimation, the Rx will decide whether the discovery request should be accepted and an acknowledgment (ACK) / non-acknowledgment (NACK) message will be sent back to the Tx.

7. If the sidelink discovery is accepted, the two sidelink counterparts will establish a security association by exchanging security-related information.

8. The result of the discovery procedure is further forwarded to the BS over the relay sensor $\#m$, taking advantage of its good cellular channel condition. If a NACK message is received by the BS, it will configure the remote sensor to the cellular TM and avoid setting up the same sidelink ends in the future.

Moreover, since both ends of a sidelink are synchronized to the same BS and they statically locate nearby each other, no additional synchronization procedure between these two users is required.

4.4.2 TM update procedure

If the physical conditions of certain sensors have been changed, e.g. the battery level of a relay sensor decreases severely and it is not sufficient to support the relay functionality anymore, and they are no longer suitable to perform the currently configured sidelink communication, TMs of the relevant sensors need to be updated. The signaling scheme to support this TM update procedure is proposed by the author and shown in Figure 4.5. Although this figure only showcases that two remote sensors are to be updated to the sidelink TM, it can be extended to the case where more sensors are assigned with sidelink TM. The signaling diagram is detailed below:

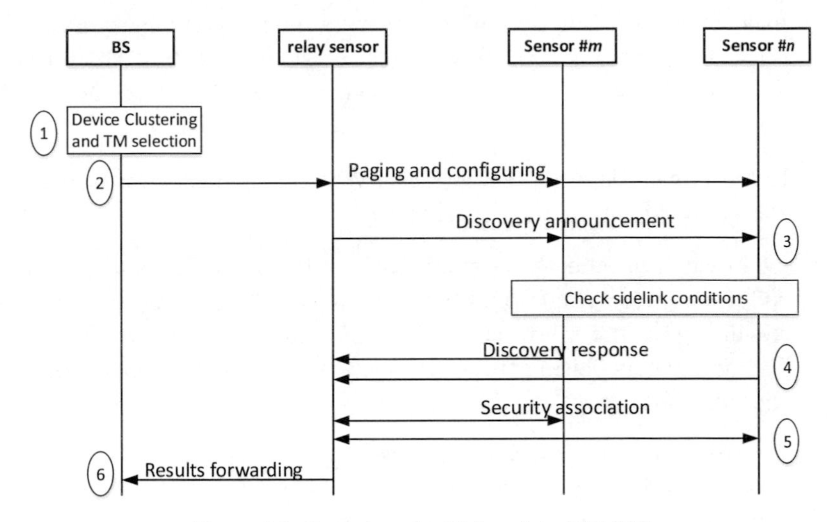

Figure 4.5: Procedure for TM update [JHLS17]

1. Based on the context information collected in real time, e.g. real-time battery level of sensors, the BS performs the context-aware TMS algorithm.

2. After obtaining the output from the context-aware device clustering and TMS algorithm, i.e. the new TMs of its serving sensors, the BS pages the relevant sensors and configures them to their new TMs. In this message, other dedicated control information is also transmitted, e.g. IDs of the sensors involved in the sidelink and the conditions for setting up a sidelink. In Figure 4.5, sensors $\#m$ and $\#n$ are configured to sidelink TM, and another sensor acts as their relay.

3. A discovery announcement message will be broadcasted by the relay sensor to the remote sensors, with the IDs of itself and also the conveyed target sensors, i.e. sensors $\#m$ and $\#n$. Reference signals are also carried in this discovery announcement message for the purpose of channel estimation.

4. Comparing the measured S-RSRP with the threshold value obtained

from Step 2, the Rxs of the discovery announcement message, i.e. sensors #m and #n, check whether the sidelink request should be accepted and an ACK/NACK message will be fed back to the relay sensor.

5. If the request is accepted, a security association between the two ends of a sidelink is then established.
6. The relay sensor further reports the results of the TM update procedure to its serving BS. If a remote sensor fails in establishing a sidelink with the relay sensor, the BS configures it to the cellular TM and avoids pairing these two sensors for a sidelink operation in the future.

Please note, since the discovery announcement message will be broadcasted by the relay sensor to multiple remote sensors in Step 3, each sensor in Step 4 can send its response by selecting a resource from the resource pool indicated by the configuration information of Step 2. In addition, once a sidelink has been successfully established in Step 5, the remote sensors in the sidelink TM should be aware of the discontinuous reception (DRX) cycle of the relay sensor in order to derive the future time slots that the relay sensor wakes up for a packet reception. Moreover, as mentioned before, in order to obtain a compromise between network adaptability and energy consumption, the TM update procedure takes place on a large time scale, with a period varying from a couple of hours to several days.

4.4.3 Uplink reports by sidelink communication

Once a sidelink is established, it will be exploited to relay the packets from the remote sensor. Figure 4.6 shows the uplink transmission procedure proposed by the author if two remote sensors in the sidelink TM are paged by the network or have data in their buffers for transmission:

1. Sensors receive the S-SIBs and DCI.
2. In the mobile terminated case, the BS will page one or multiple remote sensors to trigger their reporting procedure. In this paging message, the resource scheduled for the sidelink communication is

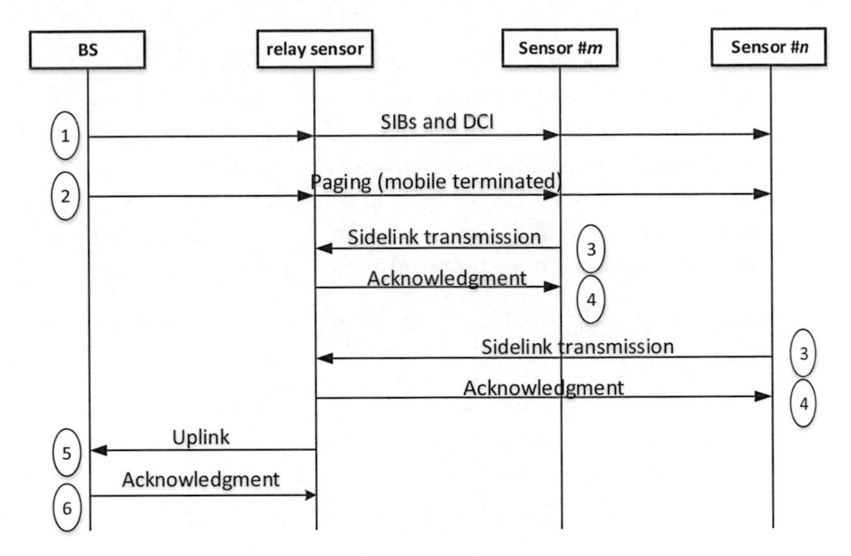

Figure 4.6: Uplink report procedure [JHLS17]

indicated. As shown in Figure 4.6, the BS will page sensor $\#m$, sensor $\#n$ and the relay sensor to upload their reports.

3. In the mobile originated case, the packets of remote sensors in the sidelink TM, i.e. sensor $\#m$ and sensor $\#n$ in Figure 4.6, will be transmitted to the BS by traveling through the relay sensor. To facilitate the sidelink transmission, there are two options to obtain the time-and-frequency resource. One alternative is that multiple remote devices need to dynamically access the relay sensor by sending different preambles whenever they need to transmit a new data packet. After that, the relay sensor is responsible for resource allocation, and the scheduling information will be sent from the relay back to the remote sensors. Another option refers to a semi-persistent resource allocation scheme where a set of periodical resources for the sidelink transmission are pre-allocated to each sensor.

4. After successfully receiving packets from remote sensors, an ACK message will be sent back from the relay sensor. Otherwise, an NACK message is transmitted back and triggers a re-transmission

procedure.

5. Furthermore, the relay sensor needs to establish a CP connection with the network and forwards its received packets to the serving BS. In addition, the BS can configure the relay sensor to compress its own report with the reports from the remote sensors and send all the reports to the BS at one time. In this case, less power is consumed at the relay sensor, since it needs to perform the CP connection establishment procedure only for once.

6. Upon receiving the uplink reports from the relay sensor, an ACK message will be sent back by the BS. Otherwise, a re-transmission procedure is triggered by an NACK message.

Please note, in order to assist the context-aware TM update procedure at the BS, the real-time sensor information, e.g. real-time battery level of sensors, should be embedded in the data reports.

4.5 Evaluation and numerical results

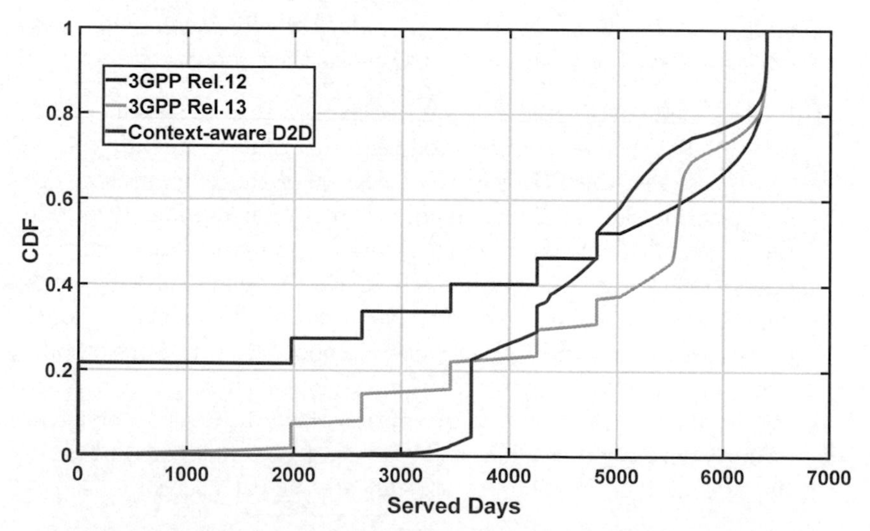

Figure 4.7: CDF of served days for all deployed mMTC devices [JHLS17]

To evaluate the proposed technique, a system-level simulator has been implemented by the author to derive the system performance w.r.t. uplink coverage and battery life of mMTC sensors. In the simulation, the author inspects the system performance in both rural and urban dense areas.
At first, one BS is deployed in the rural area with a cell radius of 2500 m [3GP16c], and 10^5 sensors are randomly distributed inside the cell radius at a height of 0.5 m. As proposed in the Long Term Evolution for Machines (LTE-M) standard [3GP13c], a maximal transmit power of 20 dBm is applied to each sensor, and a 5 Wh battery [MET16] is applied. Besides, 900 MHz is used as the carrier frequency of a Macro cell, and the RMa D2D channel model stated in Table 2.2 in Section 2.3.3.1.2 is implemented for sidelink communication. Sensors are assumed to periodically transmit data packets with a payload of 1000 bits at a frequency of one packet per 150 s. To act as a relay sensor, a relay candidate needs to have a distance to the BS larger than 1500 m. Moreover, SNR values of 10 dB and 6 dB are targeted by the open loop power control scheme to derive the transmit power of remote sensors and relay sensors, respectively. Regarding the other sensors exploiting the legacy cellular uplink transmission, their target SNR is set to 3 dB. The power consumption components provided in Table 2.8 are modeled to derive the battery life of the sensors.
The CDF of the battery life of all sensors is shown in Figure 4.7. CDF is the probability that the amount of served days is smaller than the abscissa. To provide a comparison with the legacy 4G network, the author has also given the performance of 3GPP Release 12 where D2D communication is not applied to serve mMTC services. In addition, as the concept of UE-to-network relaying has been standardized in 3GPP Release 13 for public safety scenarios, its performance in an mMTC scenario is also shown in this figure. At each time instance, if there are enough relay candidates, 100 relays will be assigned. In LTE, if a sensor experiences an uplink SNR value lower than -7 dB, no data transmission is possible and, thus, it is out of the uplink coverage [JF12]. Therefore, in Figure 4.7 more than 20% of sensors are out of network coverage and unable to transmit anything in the uplink by applying the 3GPP Release 12 technology. Please note, due to the fixed modulation and coding schemes applied in LTE, discrete steps exist in the curves.

Comparing to the 3GPP Release 12, since both the 3GPP Release 13 technology and the proposed context-aware D2D communication exploit relays, the uplink coverage can be improved. Moreover, 60% of the sensors can meet the 10-year battery life in the 3GPP Release 12, and this value is increased to 78% in the 3GPP Release 13 by utilizing the UE-to-network relaying, while the proposed approach further raises this value to 95%. As aforementioned, since relay sensors experience a high power consumption, both the 3GPP Release 13 and the proposed scheme have fewer sensors with long battery life than the 3GPP Release 12 technique.

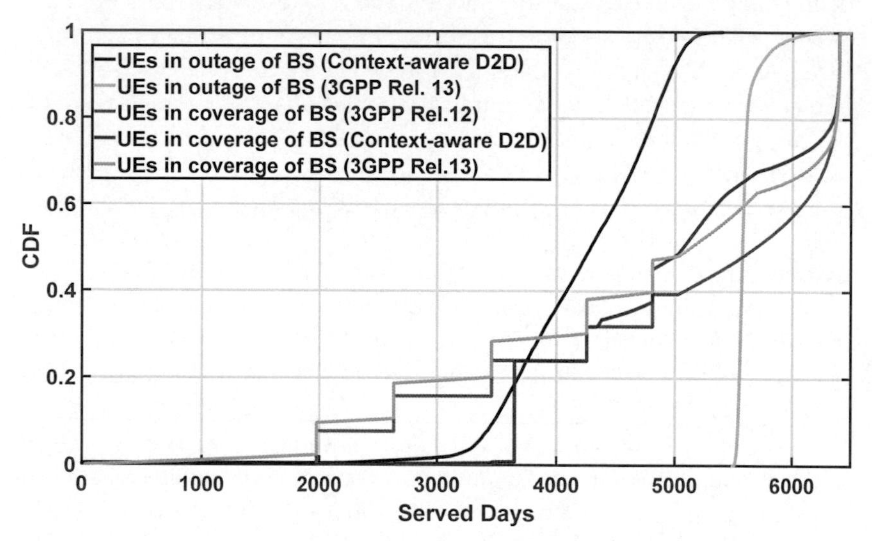

Figure 4.8: CDF of served days of mMTC devices from two different sets, i.e. in the coverage or outage of the network [JHLS17]

In order to inspect the proposed scheme in more detail, the author now categorizes the overall sensors into two different subsets based on whether a sensor is under the uplink coverage or not. The system performances of these two different subsets are separately provided in Figure 4.8. Since the BS is not aware of the battery level of its serving sensors in 3GPP Release 13, the battery of a relay sensor can run out very quickly. Thus, it can be seen from this figure that the sensors under cellular uplink coverage have worse performance in the 3GPP Release 13 compared with the sensors served by the 3GPP Release 12. However, this performance sacrifice at relay sensors in the 3GPP Release 13 can, in turn, introduce a good support for the remote sensors which are out of uplink coverage. In comparison, the proposed context-aware D2D scheme can contribute to a better performance for both the relay sensors, i.e. the dark blue curve, and the remote sensors, i.e. the black curve. Since the sensors in the network outage can be served over sidelink communication in the proposed scheme, 80% of the remote sensors observe a battery life of more than 10 years. It should be noticed that 20% of the out-of-coverage sensors can still not achieve the battery life requirement of 10 years. The reason is that relay sensors experience a high battery drain, and they are deemed by the BS as not suitable for being relays after a certain time. Therefore, with time running, there are less and less sensors suited to act as relays, and certain remote sensors cannot be served via sidelink communication anymore. In addition, the performance comparison of the sensors under network coverage is also provided in the figure. By using the proposed context-aware D2D scheme, the ratio of sensors fulfilling the battery life requirement is improved from 76%, i.e. the magenta curve, to 99%, i.e. the dark blue curve. It is worth noticing that, once a remote sensor has been served by relay sensors for more than 10 years, it will not be served via sidelink anymore, since the battery life requirement of 10 years has been already fulfilled. This point is reflected in this figure, as a step from 1% to 24% can be observed in the context-aware D2D scheme at the point of 10 years, i.e. the dark blue curve. This means that 23% sensors are served exactly for 10 years in the proposed context-aware D2D scheme.

As aforementioned, the performances of different schemes are also inspected in an urban dense scenario, i.e. Madrid grid, as modeled in Section 2.3.2.2. In this scenario, in order to achieve an inter-site distance (ISD) of 1732 m proposed in 3GPP [3GP15a, 3GP16c], a macro BS with three sectors and a cell radius of 886 m is deployed on top of the building #6 as shown in Figure 2.1. The macro BS operates on a carrier frequency of 900 MHz. In this scenario, 20 000 sensor devices are randomly distributed on the different floors of buildings. The maximal transmit power of a UE is set to 23 dBm, and an AAA battery is installed in each sensor device. Last but not least, a packet of 2000 bits with a periodicity of 24 reports per hour is implemented as the traffic model.

In Figure 4.9, the CDF of battery life is plotted w.r.t. the different clustering algorithms introduced in Section 4.3.1.2. Again, the 3GPP Release 12 technique is considered as a baseline to provide the performance of the case where D2D communication is not applied. Currently, the area size covered by each cluster is set to 40 000 m^2 which is used by the geometrical clustering algorithm to derive the number of clusters required to cover the whole area. Afterwards, in order to achieve a fair comparison among different clustering algorithms, the same number of clusters is used in the other clustering algorithms. In addition, the target SNR threshold to successfully set up a sidelink is 2 dB. Part of the CDF plot is zoomed in and shown at the bottom of Figure 4.9, since it is the most interesting part for our inspection. Looking at the point where the served days is equal to zero, it can be seen that 14% of the sensors are not served by the network at all and, therefore, they are out of the uplink coverage by the legacy 4G technology. This ratio can be reduced to 2% if sidelink communication is enabled. Thus, the network uplink coverage is improved from 86% to 98%. Furthermore, looking at the point where the served days is equal to ten years, 80.5% of the sensors can fulfill the battery life requirement of ten years in the LTE system, while the proposed D2D communication facilitate 90% of the sensors to achieve the 10-years battery life. Moreover, the performance deviation of the four different clustering algorithms is within a range of 2%.

In Figure 4.10, the cluster size for the geometrical clustering algorithm is shrunk to 2500 m^2 to derive the number of required clusters, which makes the difference compared with Figure 4.9. As we can see from the figure, the reduced clustering size leads to a more significant performance deviation for the four different clustering algorithms, compared with the case before. Due to a smaller cluster size, more clusters are required to cover the whole area, and each cluster comprises a smaller number of UEs. Thus, there is a lower probability of having a feasible relay candidate in each cluster. This explains why the uplink coverages of the three clustering algorithms, i.e. labeled as “Distance based clustering”, “K-means clustering” and “Geometrical clustering”, are decreased from 98% in Figure 4.9 to 96% in Figure 4.10 by looking at the point where the served days is equal to zero. As an exception, the availability of the “Distance+CSI based clustering” algorithm is improved to 98.5% due to the fact that sensors with good channel conditions are selected as centroids in the initial step of the clustering algorithm. Besides, since only intra-cluster D2D communication is allowed, the sidelink pathloss value is statistically smaller due to a reduced cluster size. Thus, a smaller pathloss value together with the open loop power control scheme can contribute to a lower energy consumption for each packet transmission. This illustrates why all the clustering algorithms can provide more sensors meeting the 10-years battery life in Figure 4.10 than in Figure 4.9. In particular, the "Distance+CSI based clustering" algorithm provides the best performance w.r.t. the ratio of sensors fulfilling the battery life requirement, and the number is improved from 80.5% in 4G to almost 95% by looking at the point where the served days is equal to ten years.

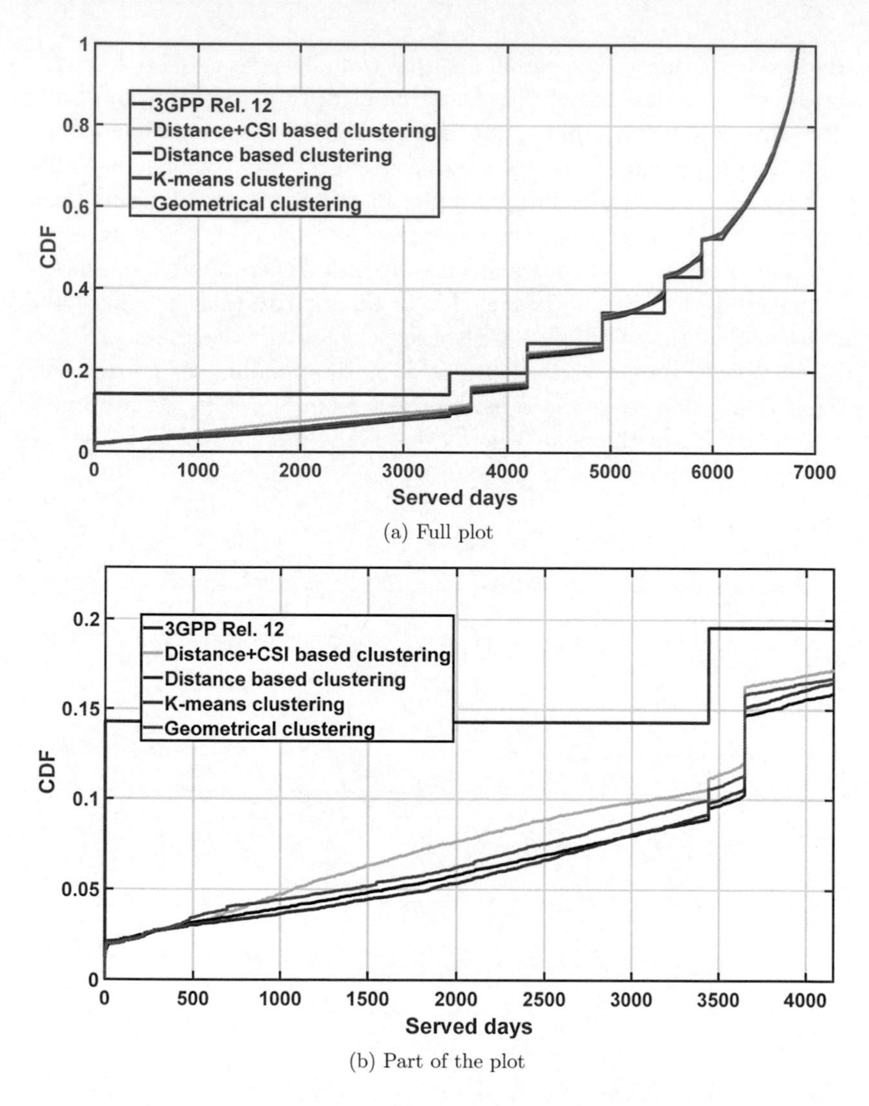

Figure 4.9: CDF of served days, $A_{\text{sector}} = 40000\ \text{m}^2$ [JLS17]

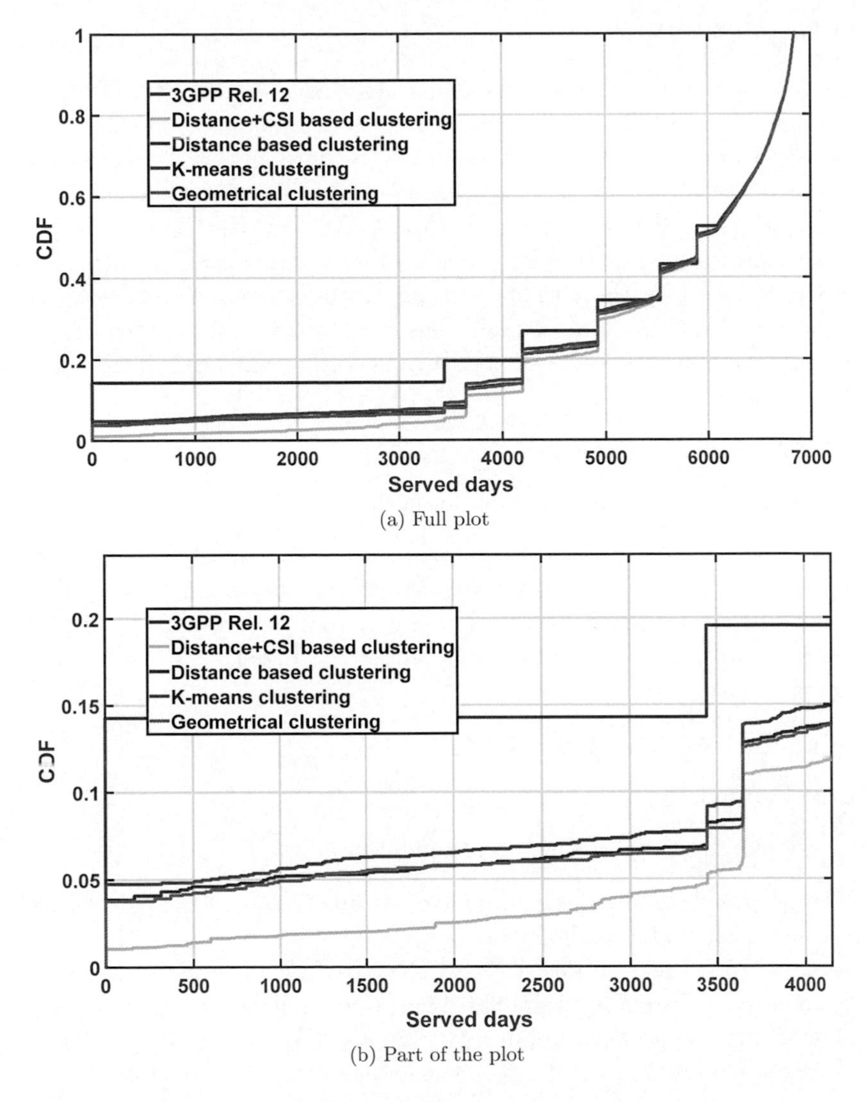

(a) Full plot

(b) Part of the plot

Figure 4.10: CDF of served days, $A_{\text{sector}} = 2500\ \text{m}^2$ [JLS17]

4.6 Summary

In this chapter, the author has shown the technical challenges to support the emerging mMTC services with the legacy cellular technology. In order to improve the network coverage in the uplink and prolong the battery life of the remote sensors with harsh radio channel conditions, some sensors have been configured to act as relays to forward packets from remote sensors to the BS. In this sense, sidelink communication has been exploited by the author to facilitate the communication between a relay and its associated remote sensors. Moreover, the context-aware sidelink communication developed by the author has been divided into two steps:

- clustering of sensors;
- transmission mode selection for each sensor.

Since the network should be equipped with the capability to cope with different scenarios, the author has developed different clustering algorithms for both the rural and urban areas. In order to support the proposed technology, three critical signaling procedures have been designed by the author:

- Initial attachment of sensors;
- TM update procedure;
- Uplink reports by sidelink communication.

The proposed signaling schemes have the ability to collect the required context information without a heavy signaling overhead.
Last but not least, to derive the system performance of the proposed technology, a system-level simulator has been implemented by the author to evaluate the network uplink coverage and the battery life of mMTC devices. Compared with the legacy 4G technology, the numerical results have demonstrated a significant performance gain by using the proposed context-aware sidelink communication to enhance the mMTC services. For instance, in the dense urban area, the network uplink coverage can be improved from 86% to 98.5%, and the amount of sensors fulfilling the battery life requirement is increased from 80.5% to 95%.

Chapter 5
Network-controlled V2X Communication

5.1 Introduction

The current 4G mobile radio system offers a good connectivity for most of the cases. However, in areas with poor coverage, under excessive interference or where network resources are overloaded, the reliability of wireless links cannot be guaranteed [MET14a]. Moreover, the legacy 4G system is not sufficient to deliver a solution for the mission-critical communications characterized by the need for ultra-high reliability and a low latency. In the years beyond 2020, the new generation of wireless systems will be designed to offer solutions for new services requiring a high degree of reliability and availability in terms of data rate, latency and other QoS parameters [MET13a]. For instance, compared with the legacy 4G system, the main technical goals of URLLC in 5G are [NGM15]

- 10 to 100 times higher number of connected devices: URLLC by a reliable service decomposition providing mechanisms for an increased number of users.
- 10 times longer battery life for low power devices: URLLC proving beneficial in disaster applications when using low power devices.
- 5 times reduced E2E latency: URLLC providing reduced E2E latency in real-time applications.

As a typical use case for URLLC, V2X communication has attracted an intensive research work in 5G system design to meet the demand for vehicular communication aiming at an improvement of traffic safety and efficiency. The V2X communication taking place between two nearby traffic participants, e.g. vehicles and pedestrians, can facilitate a local information exchange in advanced and automated driving applications, and it has strict requirements on latency and transmission reliability. As mentioned previously in Chapter 1, V2X communication includes different communication profiles such as V2I, V2P, V2N and V2V as shown in Figure 5.1.
To technically support V2X communication, one of the solutions considered by 3GPP is to transmit data packets through the cellular network infrastructure via the LTE-Uu air interface [JWHS17, WWFD18,

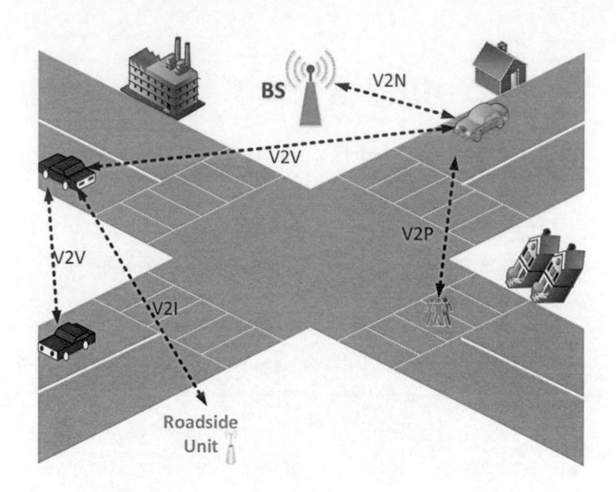

Figure 5.1: Automotive V2X communication

SSPR16, 3GP16a] which is used for the communication between a BS and its serving mobile users in the RAN. This solution relies on the technical enhancements of the LTE-Uu radio interface, e.g. a shorter TTI duration, together with the evolution of the network architecture, e.g. applying mobile edge computing (MEC). In this approach, a BS or an eNodeB-type Roadside Unit (RSU) can receive the packets transmitted in the uplink and then forward them to the relevant Rxs in the downlink direction. In V2X communication, the RSU is a computing device locating on the roadside and providing connectivity support to passing vehicles. Moreover, an RSU can either be a UE-type RSU connected to the BS with a backhaul link or even act as a BS, i.e. as an eNodeB-type RSU [3GP18a]. A UE-type RSU is only a normal UE from radio perspective and, thus, its connection to another UE will be via the PC5 interface. Comparing to that, an eNodeB-type RSU has the functionalities of an eNodeB, and it uses the Uu interface to communicate with another UE. Therefore, an evaluation of the feasibility of using the LTE-Uu interface to offer V2X services should be carried out, which is one of the targets in this chapter.
Additionally, a lot of work in the literature focuses on the exploitation of a direct V2X communication [JLWS17, MMG17, Vin12, ACC+13,

3GP17d], where a data packet is directly transmitted from a Tx to its nearby Rx(s) without going through the network infrastructure. For instance, 3GPP proposes the PC5 interface [3GP17d] to facilitate the direct communication between the two ends of a V2X communication [3GP18f]. Alternatively, the European Telecommunication Standards Institute (ETSI) proposes the Cooperative Intelligent Transport Systems (C-ITS) to apply the IEEE 802.11p protocol as the air interface for the direct V2X communication [IEE10]. One of the advantages of applying the direct V2X communication is that the packet E2E latency in the UP can be efficiently reduced, since the network infrastructure is not involved in the data transmission. It is worth noticing that, when a direct V2X communication takes place among a set of users over the PC5 interface or the IEEE 802.11p protocol, it is performed as a point-to-multi-point (P2MP) multicast transmission in the radio layers. In this way, all relevant traffic participants can get the information of the Tx, if the received signal has a good quality.
In this chapter, the author first carries out an inspection of applying the air interfaces LTE-Uu and PC5 in the cellular system to enable V2X communication. In Section 5.2, the V2X communication applies the LTE-Uu interface to transmit data packets through network infrastructures. Following that, the direct V2X communication via the PC5 interface is detailed in Section 5.3. Moreover, to improve packet transmission range and improve communication reliability, a two-hop direct V2X communication scheme is proposed in Section 5.4, and a multi-RAT concept is presented in Section 5.5 to obtain a diversity gain. Finally, this chapter is summarized in Section 5.6.

5.2 Cellular V2X communication through network infrastructures

5.2.1 System model and evaluation methodology

In order to improve traffic efficiency and help drivers to avoid accidents, the C-ITS relying on a timely and reliable information exchange among nearby traffic participants can be applied [ETS13]. The purpose of this information exchange procedure is to share the status information, e.g. position, velocity and acceleration, of a traffic participant with others.

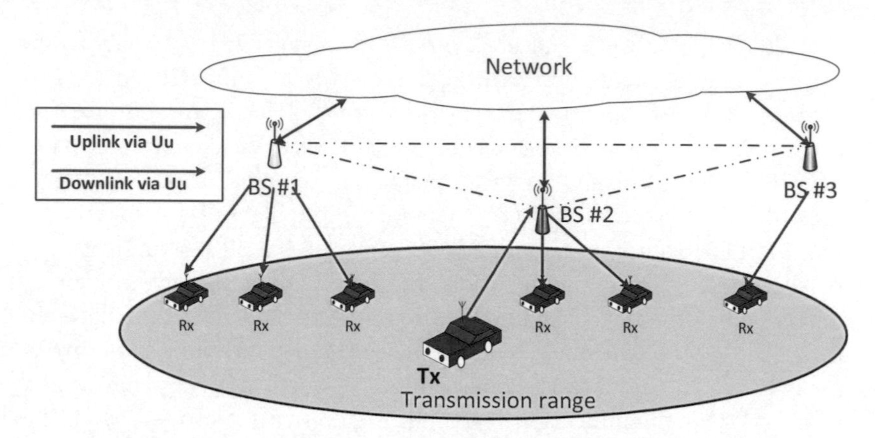

Figure 5.2: V2X communication through the cellular network infrastructure [JWHS18a]

Compared with the human-driven services provided by the LTE network, the main technical challenges here refer to a low packet E2E latency with ultra-high reliability. If a V2X packet is transmitted from one UE to another UE over the LTE-Uu interface, it refers to a communication process shown in Figure 5.2, and the packet E2E transmission process can be decomposed into three steps:

1. A packet generated at a vehicle is transmitted to the serving BS #2 via the LTE-Uu interface in the uplink;
2. the BS #2 sends the packet to the serving BSs #1 and #3 of the relevant Rxs through network infrastructures;
3. the packet is further transmitted to each Rx from its serving BS #1, #2 or #3, respectively, via the LTE-Uu interface in the downlink.

As aforementioned, in order to assist advanced driving, a V2X packet is intended to be received by a group of traffic participants within a certain radius of the Tx. As the group of target Rxs might be served by different BSs, the packet can be sent to multiple BSs in Step 2. In Figure 5.2, the network needs to transmit the packets from the BS #2 to the BSs #1 and #3 in order to distribute the received packets to all relevant Rxs. Corresponding to the steps listed above, the UP E2E latency representing the one way transmission time of a packet between its generation at the Tx and its successful arrival at the Rx comprises three components:

- Uplink latency: The time difference between the generation of one packet at the Tx and its successful reception at the BS #2 in the uplink.
- Propagation latency among BSs: The time for the serving BS #2 of the Tx to successfully transmit the packet to the serving BSs #1 and #3, respectively, of the relevant Rxs.
- Downlink latency: The time difference between the packet successfully arriving at the serving BSs #1, #2 and #3 of the Rxs and its successful arrival at the Rx.

Below, these three latency components will be discussed in more detail.

5.2.1.1 Uplink latency

As it can be seen from Figure 5.2, the uplink transmission refers to a P2P transmission. In the following, the critical aspects to apply the LTE-Uu uplink for V2X communication will be highlighted.

5.2.1.1.1 Resource scheduling Before an uplink transmission starts, a V2X Tx needs to query the BS #2 for the time-and-frequency resource for its transmission. So far, 3GPP has developed two scheduling schemes to assign a resource for V2X communication, i.e. dynamic scheduling and semi-persistent scheduling (SPS).
In the dynamic scheduling scheme, a scheduling request (SR) message needs to be sent to the BS, once a packet arrives at the buffer of a Tx. Following that, a resource will be scheduled by the BS, and the corresponding configuration information will be sent back to the Tx in the downlink control information (DCI). Afterwards, the Tx can start its transmission by using the allocated resource. However, this dynamic scheduling approach is not always efficient for V2X communication.
A big portion of the V2X traffic refers to the Cooperative Awareness Messages (CAMs) which are a set of periodical transmitted messages for providing information of the Tx, e.g. its presence, position, temperature, and basic status [SPMS13, SPMS14]. In a real system, the transmission frequency of the CAMs can be quite high. For instance, a frequency up to 40 Hz is required in a vehicular platooning scenario where a group of vehicles operate in a closely linked manner and the vehicles move

like a train [3GP17e]. In this case, a high signaling overhead is to be expected with the dynamic scheduling approach since the Tx needs to request a resource for each packet. Compared to that, the SPS scheme developed by 3GPP enables the network to assign a set of time-periodic resources to a V2X Tx. After obtaining the SPS configuration information, the Tx can periodically transmit its data without further triggering a resource request procedure until the SPS configuration information expires. Thus, compared with the dynamic scheduling, the SPS provides a better support for the CAMs due to its less signaling overhead.

5.2.1.1.2 HARQ transmission As aforementioned, the V2X communication standardized by 3GPP refers to a P2P transmission in the uplink and, thus, the radio link control (RLC) layer operates in the acknowledged mode (AM), where an ACK/NACK message will be sent back from the Rx to the Tx to notify the packet reception status. In case a packet is not successfully delivered in the uplink, an NACK message will be sent to the Rx to trigger an HARQ procedure. It should be noticed that the minimal time interval between the end of a packet transmission and the start of its retransmission is set to be 7 ms in the LTE.

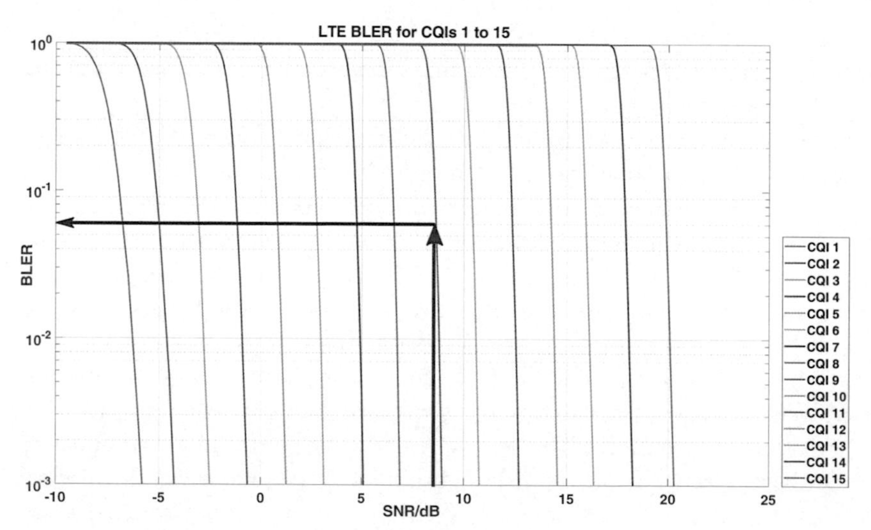

Figure 5.3: Mapping from SNR to block error rate [IWR10]

5.2.1.1.3 Modulation and coding scheme In order to provide a good compromise between robustness and spectral efficiency, multiple MCSs have been designed by 3GPP. For instance, an MCS with a low spectral efficiency is able to preserve a good robustness and, therefore, it is used to counter the effect of a bad channel condition. For instance, to offer the services demanding a high data rate without a strict latency requirement, a Tx can select the MSC scheme, which is able to provide a high throughput. In this case, retransmission will be needed to correct the block error. Figure 5.3 shows an example regarding how a Tx selects the MCS for its radio link, based on the SNR value and the network-configured BLER threshold in the LTE, e.g. 10%. In this figure, the Tx selects the MCS that has the highest spectral efficiency while providing a BLER lower than the configured threshold. However, under the scope of V2X communication, the ultra-high reliability and low latency requirements pose the demand to select the MCS guaranteeing an ultra-low BLER. In this manner, retransmission can be avoided.
In Procedure 1, the author proposes the approach to evaluating uplink latency by tracking each packet from its creation until its successful arrival at the serving BS. In case a packet fails in the uplink transmission, the target Rxs cannot receive this packet in the downlink, and the packet E2E latency is considered infinite. Please note that, since the UP E2E latency is the main focus here, the CP functions are not considered in the procedure, e.g. system information acquisition, mobility control and state transition. However, it should be noticed that before a UE communicates with the BS, a CP connection with the network should be established. This CP connection establishment procedure requires the UE to execute several steps [3GP18f], such as synchronizing with the BS, obtaining the network configuration by receiving the master information block (MIB) and SIBs, and performing the random access (RA) procedure.

5.2.1.2 Propagation latency between BSs

In the situation shown in Figure 5.2, a packet successfully received in the uplink needs to be transmitted to other relevant BSs before the downlink transmission can take place. This transmission is a P2MP transmission in the domain of a core network, and the multimedia broadcast multicast services (MBMS) standardized by 3GPP can serve as an efficient

Procedure 1 Evaluation of uplink latency [JWHS18a]

1: A V2X packet is generated at the Tx. In a period of 100 ms, packets generation time among different Txs has a uniform distribution.
2: Perform transport block cyclic redundancy check (CRC) attachment and block segmentation if it has a size greater than 6144 bits.
3: Decide coding and modulation scheme w.r.t. SINR value of each receiver, which is generated by the simulator.
4: BLER is derived from the SINR value of each receiver w.r.t. the coding and modulation scheme selected from Step 3.
5: Round robin scheduler is used to determine how many resource blocks are allocated to each uplink packet.
6: Uplink packet starts to be transmitted to the serving BS.
7: If a packet is not successfully received, w.r.t. BLER of Step 4, HARQ retransmission will be initialized and we inspect on whether HARQ retransmission is possible and successful.
8: Once a packet is successfully received by the serving BS, the timing instance when this packet is received is recorded. If the packet transmission cannot be successful, the packet delay is considered infinity.

interface specification to support the multicast transmission [3GP17c]. Compared with the system architecture used for P2P transmission, two additional entities are deployed in the network architecture for MBMS, i.e. the Broadcast Multicast - Service Center (BM-SC) and the MBMS gateway (MBMS-GW) [3GP17b]. The BM-SC takes the responsibility to manage the MBMS service-related information, e.g. mapping the service information to the QoS parameters, while the MBMS-GW delivers MBMS traffics to multiple cell sites. In order to meet the low latency requirement in V2X communication, 3GPP has proposed to enhance the MBMS architecture by localizing the functional entities at the edge of the RAN [3GP16a]. In this work, as the focus is on the RAN, it is assumed that the architecture for MBMS has been enhanced by localizing the functional entities, and a propagation latency between two different BSs is 1 ms [MET14b].

5.2.1.3 Downlink latency

After the packet is successfully delivered to the serving BS of the Rx, the BS will reserve some time-and-frequency resources for the downlink transmission. So far, two transmission modes exist in the cellular downlink transmission. As the first option, a unicast transmission mode can be utilized, where the packet is delivered to the receivers in a P2P manner. Thus, in case that N Rxs require a packet, the BS needs to transmit the packet in the downlink N times and each time target at a specific Rx. Please note that in the unicast transmission mode, the MCS for each Rx can be independently selected by the BS according to the CSI. Moreover, the RLC layer in the unicast transmission mode can operate in the acknowledged mode (AM) meaning that an ACK/NACK message needs to be sent back from each Rx to its serving BS, depending on whether a packet is successfully received or not. To evaluate the packet downlink transmission latency in the unicast mode, Procedure 2 can be applied.

Procedure 2 Evaluation of downlink latency with unicast transmission [JWHS18a]

1: Only packets successful received by the BSs in the uplink will be transmitted in the downlink.
2: A packet arrives at the BS. Its arriving time at the serving BS of a Rx is equal to the arriving time of the packet in the uplink, which is calculated in Procedure 1, plus the propagation latency between the BSs, i.e. 1 ms in this thesis.
3: Perform transport block CRC attachment and code block segmentation on each packet.
4: Decide coding and modulation scheme w.r.t. SINR value of each receiver.
5: BLER is derived from SINR value of each receiver w.r.t. the coding and modulation scheme selected in Step 4.
6: The BS allocates the time and frequency resource to the most recently received packets. In case multiple packets are ready to be transmitted simultaneously, a round robin scheduler is used to decide how many frequency resource blocks are allocated to each downlink packet.
7: Downlink packet starts to be transmitted to the receiver.
8: If a packet is not received correctly w.r.t. BLER of Step 5, HARQ retransmission will be triggered and we inspect whether the HARQ retransmission is possible and successful.
9: Once the packet is successfully received by the receiver, the timing instance is recorded. If the packet transmission cannot be successful, the packet delay is considered infinite.

The second alternative in the downlink transmission corresponds to a multicast mode which supports a P2MP radio transmission. As a specification of the LTE-Uu air interface, the MBMS can also be applied to enable the multicast mode in the RAN. Since V2X communication targets at distributing the information of one vehicle to other nearby users, the multicast transmission mode in the LTE downlink can multicast a packet to N Rxs simultaneously. However, the RLC layer will operate in the unacknowledged mode (UM) to support the multicast transmission. It means that Rxs do not send any feedback message to the Tx, no matter whether the receptions are successful or not. Therefore, in order to successfully transmit a packet to all the Rxs, the BS should provide a good robustness to the Rx with the worst reception condition, e.g. the lowest SINR value. As an efficient approach to improve reliability, the BS can autonomously repeat the packet transmission if there is enough available resource. In this case, if a packet is not successfully received from the first transmission trial, the Rx can try to receive the retransmitted packet and perform a maximal ratio combining (MRC). To evaluate the downlink latency in the multicast transmission mode, Procedure 3 indicates the methodology used in this thesis.

Procedure 3 Evaluation of downlink latency with multicast transmission [JWHS18a]

1: Only packets successfully received by the BSs in the uplink will be transmitted in the downlink.
2: A packet arrives at the BS. Its arriving time at the serving BS of a Rx is equal to the arriving time of the packet in the uplink, which is calculated in Procedure 1, plus the propagation latency between the BSs, i.e. 1 ms in this thesis.
3: Perform transport block CRC attachment and code block segmentation on each packet.
4: Decide coding and modulation scheme based on the network condition, taking into account of the overall traffic volume and the overall available bandwidth.
5: The BS allocates its time and frequency resource to the most recently received packets. In case multiple packets are ready to be transmitted simultaneously, a round robin scheduler is used to determine how many frequency resource blocks are allocated to each downlink packet.
6: Within the allocated resource, BS multicasts the packet. In case with extra available resource, the packet transmission will be repeated in order to fully utilize the available resource.
7: BLER is derived from the SINR value of each receiver. Based on the BLER value, whether a packet transmission is successful or not can be derived.
8: In case a packet reception fails and a repetition of this transmission is applied, an HARQ process referring to the chase combining is carried out at the receiver. The HARQ process will introduce a new effective SINR value and correspondingly a new BLER. With this new BLER value, Step 7 is repeated.
9: If a packet is successfully received, the reception timing instance will be recorded. If a packet is not successfully received in the allocated time resource, the packet will be discarded by the BS and considered having an infinite delay.

Table 5.1: UP latency parameters [JWHS17, 3GP17a]

Description	Value
UE processing delay	1 ms
Frame alignment	0.5ms
TTI for uplink (UL)/downlink (DL) packets	packet specific
HARQ retransmission	7 ms
eNB processing delay	1 ms
Packet exchange between eNBs	1 ms

5.2.2 Numerical results

The Madrid grid model introduced in Section 2.3.2.2 is used to model a dense urban scenario, and the macro BS with three sectors shown in Figure 2.1 is deployed to serve the V2X communication. Moreover, the LTE-Uu interface operates on a carrier frequency of 800 MHz with a variable bandwidth dedicated to V2X communication in order to investigate the impact of the spectrum resource. An antenna is installed on each vehicle with a height of 1.5 m and a constant spectral transmit power density of -46 dBm/Hz corresponding to a transmit power of 24 dBm in a 10 MHz bandwidth. In addition, vehicles are deployed on the roads with a density of 1000 users/(km)^2, and the Rxs located within a range of 200 m of the Tx should receive its data packets.
Moreover, the UP latency parameters proposed by 3GPP [3GP17a] are applied as listed in Table 5.1. Please note that the minimum delay between the end of a packet transmission and the start of its retransmission is 7 ms in the LTE-FDD unicast transmission mode [HT09], and it should be taken into account to compute the UP E2E latency when an HARQ retransmission is triggered by an NACK feedback message. In addition, the required number of TTIs for one packet transmission depends on the spectrum efficiency of the applied MCS and also on the allocated bandwidth.
The CDF plot of the packet E2E latency w.r.t. different bandwidths is derived by the author as shown in Figure 5.4. To be specific, a series of bandwidths ranging from 10 MHz/10 MHz to 100 MHz/100 MHz for uplink/downlink have been applied to demonstrate the system performance. The marks in this figure represent the discontinuous values of

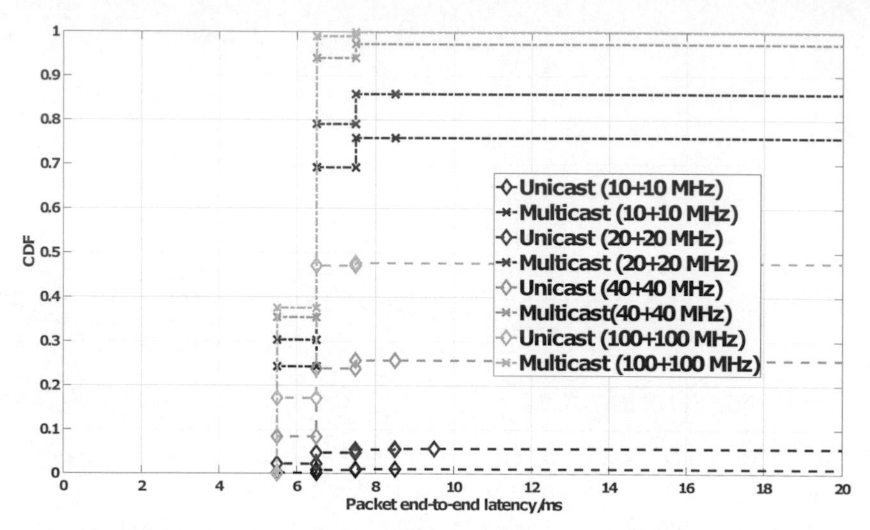

Figure 5.4: CDF plot of the E2E latency for performance comparison between Unicast and Multicast (Spectral efficiency of multicast = 0.877 bit/Hz) [JWHS17]

the packet E2E latency based on the latency assumptions from Table 5.1. Moreover, the CDF curves in this figure do not converge to the value 1 because some packets are not successfully delivered to the Rxs and, therefore, these packets are considered having an infinite E2E latency. In addition, both the unicast and multicast transmission modes in the downlink are implemented to compare their performances. Besides the CDF plot, the average E2E latency and the PRRs are also provided by the author in Table 5.2. As mentioned in Section 5.2.1, the V2X communication refers to a P2MP transmission, where the uplink is a P2P transmission and the downlink is a P2MP transmission. Thus, if a unicast transmission is applied for the downlink, a larger bandwidth is required in the downlink than in the uplink. From both Figure 5.4 and Table 5.2 it can be seen that a large bandwidth is required by the unicast transmission mode in the downlink to cope with the considered user density, and that is the reason why an increased bandwidth can significantly improve the performance. In comparison, the multicast transmission mode applying an MCS with a spectral efficiency of 0.877 bit/Hz obviously outperforms the unicast transmission mode with the same amount of spectral resource.

Table 5.2: Average latency and PRR [JWHS17]

Bandwidth (UL+DL)/MHz	Mean latency by unicast	PRR by unicast	Mean latency by multicast	PRR by multicast
10+10	6.649 ms	1.07%	6.269 ms	75.9%
20+20	6.323 ms	5.75%	6.227 ms	85.91%
40+40	6.246 ms	25.75%	6.168 ms	97.26%
100+100	6.151 ms	47.68%	6.133 ms	99.91%

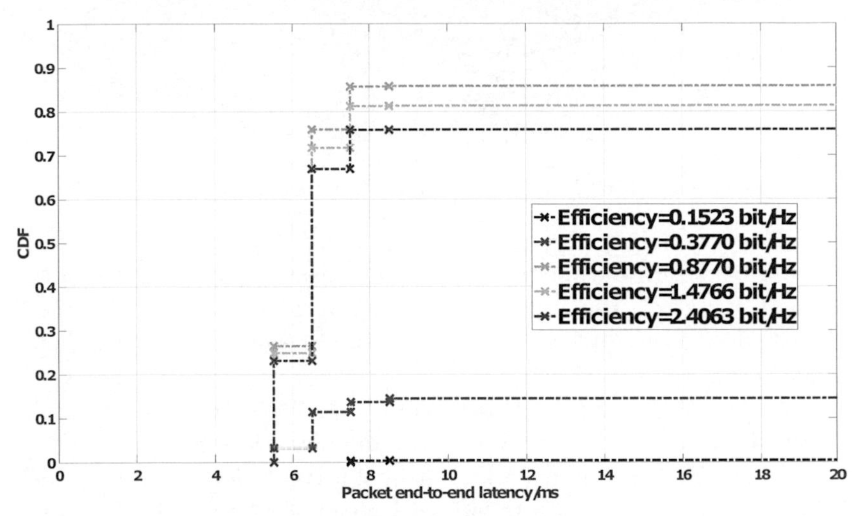

Figure 5.5: CDF of packet E2E latency by Multicast [JWHS17]

In Figure 5.5, the performance of the multicast transmission mode with different MCSs is given. Again, the marks in this figure show the discrete packet E2E latency values and the curves do not converge to value 1 because some packets fail in the transmission. Theoretically, an MCS with lower efficiency requires more resources in the downlink. Thus, due to the lack of resources, the MCS with the efficiency of 0.877 bit/Hz outperforms the two MCSs with the efficiency of 0.1523 bit/Hz and 0.377 bit/Hz. On the other hand, the MCSs with higher efficiency, e.g. the ones with the efficiencies 1.4766 bit/Hz and 2.4063 bit/Hz in Figure 5.5, are not robust for the Rxs experiencing low SINR values. Many Rxs fail in the packet reception with these high-efficiency MCSs. Thus, a com-

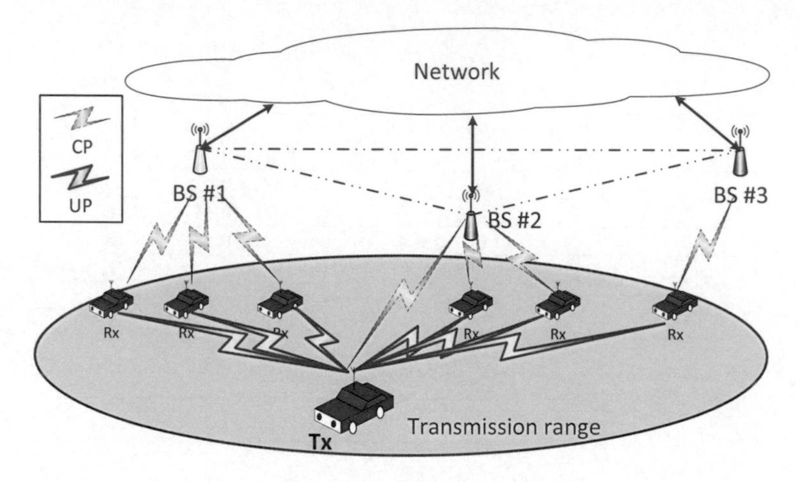

Figure 5.6: Direct V2X communication with network assistance [JWHS18a]

promise between robustness and efficiency needs to be achieved, when the network selects the MCS for the multicast transmission. The selection should take account of both the real-time system load and the radio conditions of the Rxs.

Please note that, unicast transmission may outperform the multicast transmission in certain scenarios. For example, if the V2X application layer at a V2X UE requires a point-to-point transmission with another V2X UE, it is better to utilize unicast instead of multicast in the downlink due to a better efficiency. In addition, if there is a small density of V2X UEs under the network coverage, e.g. in the middle night, unicast can also outperform the multicast transmission since multicast in the downlink does not support the acknowledged mode.

5.3 Direct V2X communication via the PC5 interface

Since V2X communication refers to a local information exchange, a direct V2X communication where a Tx directly transmits its data packets to the surrounding Rxs has been proposed in 3GPP [3GP17d] as shown in Figure 5.6. In this section, the standardized direct V2X communi-

cation by 3GPP will be introduced. In order to enable the network to properly control and configure the direct V2X communication, the UEs need to establish a connection with the network in CP by using the LTE-Uu interface in Figure 5.6. The CP functions provide a coordination for the data transmission in the UP, e.g. configuring the transmit resource for the packet transmission via sidelink communication.
As aforementioned, a direct link between the two UEs is referred as sidelink in 3GPP and the applied air interface is the PC5 interface [3GP17d]. Up to 3GPP Release 15, V2X sidelink communication is connection-less which means that no Radio Resource Control (RRC) connection exists between the two ends of a direct link. In this sense, a data packet is multicasted by a V2X Tx to all of its surrounding Rxs, and an Rx can discard the received packet if it locally identifies the packet as irrelevant.
So far, 3GPP has introduced two V2X sidelink transmission modes [3GP18e] to assign radio resources in the LTE, e.g. sidelink transmission mode 3 and sidelink transmission mode 4. In the V2X sidelink transmission mode 3, the resource to transmit V2X packets needs to be scheduled by the BS. In order to obtain the scheduling information, a V2X Tx should first enter the RRC_CONNECTED state, and then an scheduling request message will be sent to the BS. After the BS receives the SR message, it will reserve certain resources and send this information as DCI to the V2X Tx. In order to support the periodical transmitted V2X packets in an efficient manner, the SPS introduced in Section 5.2.1.1.1 can also be exploited for sidelink resource allocation. Moreover, the BS can collect relevant UE context information, e.g. traffic pattern and geometrical information, to assist its resource allocation. In the other case, the V2X sidelink transmission mode 4 has been introduced to serve the V2X Txs that are out of the cellular network coverage or in the RRC_IDLE state. In this mode, the information of the resource pools is either broadcasted in the system information block (SIB) [3GP18b] or pre-configured at the V2X UEs. In this way, the V2X Tx can autonomously select the transmission resource from the resource pools. In V2X communication, the mutual interference among different sidelinks transmitting over the same resource can deteriorate the communication reliability and, thus, it needs to be controlled properly by the network [MVSAT+18]. In order to achieve this purpose, the SIB 21 can

be used by the BS to broadcast the information regarding how to map from a geometrical zone to its associated transmission resource pools [3GP18b] as shown in Figure 5.7. In this figure, the world is divided into different geographical zones, and a V2X UE needs to determine its located geographical zone by performing a modulo operation using its geo-location and the information provided by the SIB 21, e.g. the length and width of a zone, the number of zones in length, the number of zones in width, and a reference point with the coordinates (0, 0). The length and width of a zone are represented by Z_{L} and Z_{W} in this thesis, and the number of zones in length and in width by NZ_{L} and NZ_{W}, respectively. The modulo operation finds the remainder after division of one number by another, and it leads to the periodical appearance of the geographical zones in Figure 5.7. For instance, in Figure 5.7, we assume that both Z_{L} and Z_{W} are set to 100 m, and there are two zones in length and two zone in width, i.e. $NZ_{\mathrm{L}} = NZ_{\mathrm{W}} = 2$. In this case, we can see that the rectangular area from (0, 0) to (100, 100) covers the area of Zone 1 in Figure 5.7. And the rectangular area from (100, 0) to (200, 100) covers Zone 2. The rectangular area from (0, 100) to (100, 200) covers Zone 3, and the area from (100, 100) to (200, 200) covers Zone 4. As one example, if the coordinates (x, y) of a vehicle are (350, 250), the vehicle first performs the modulo operation, as

$$x' = x \text{ modulo } (NZ_{\mathrm{L}} \times Z_{\mathrm{L}}) = 350 \text{ modulo } (2 \times 100) = 150, \tag{5.1}$$

and

$$y' = y \text{ modulo } (NZ_{\mathrm{W}} \times Z_{\mathrm{W}}) = 250 \text{ modulo } (2 \times 100) = 50. \tag{5.2}$$

After obtaining the results from the modulo operation, the vehicle can map the coordinates (x', y') of (150, 50) to the specific zone that covers this location, i.e. Zone 2. In addition, to reduce the inter-zone interference and avoid the packets from different Txs colliding at one Rx, the network assigns non-overlapping transmission resource pools to any two neighboring zones. In Figure 5.7, the different resource pools are indicated by different colors. After obtaining the SIB 21, a V2X Tx in mode 4 can either randomly select a resource or perform a channel-sensing based resource selection [3GP18b]. In case a V2X Tx is out of the network coverage, it cannot receive the SIB 21, and thus the mapping information from the geographical zone to the resource pools needs to be pre-configured at the UEs.

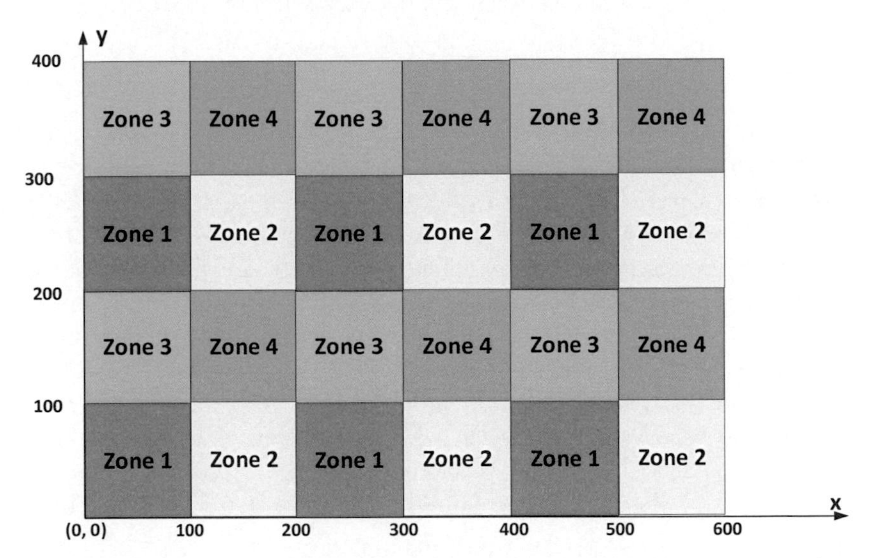

Figure 5.7: Mapping from geo-location to geographical zone and the corresponding transmission resource pool [JWHS18a]

Up to 3GPP Release 15, the transmission over the PC5 does not involve any radio layer feedback from an Rx to its corresponding Tx. Therefore, to achieve high reliability, a blind retransmission scheme can be used to multicast identical copies of the original transmission. In the blind retransmission scheme, the Rxs can perform the MRC based on all received copies in order to successfully decode a data packet. Moreover, the network needs to configure an appropriate MCS for the multicast transmission over the PC5 by taking account of both system capacity and link robustness. The required system capacity C in bits/s can be calculated as

$$C = S \times R_{\mathrm{P}} \times N_{\mathrm{v}}, \tag{5.3}$$

where N_{v} stands for the number of vehicles under the coverage of the BS, and S shows the per packet size in bits. Moreover, R_{P} is the amount of transmitted packets per second. The values for S and R_{P} used in this thesis for the V2X communication are stated in Table 2.7 in Section 2.3.5, and the value for N_v are derived based on the concrete simulation parameters, e.g. user density and the coverage area of a cell. Thus, the equation should be fulfilled when the network selects the MCS,

$$SE \geq \frac{C}{BW}, \tag{5.4}$$

where SE and BW represent the spectral efficiency of the MCS and the system bandwidth, respectively. The spectral efficiency refers to the gross bitrate divided by the bandwidth. Please note that Equation (5.4) provides the lower bound of the spectral efficiency of the selected MCS. If the selected MCS has a spectral efficiency higher than the lower bound, all the packets can be scheduled for transmission and addition resources can be available for other purposes, e.g. packet retransmission. However, if the spectral efficiency of the selected MCS does not meet the condition in Equation (5.4), some packets cannot be scheduled for transmission. Additionally, since an MCS with a higher spectral efficiency provides a worse robustness, the network should apply the MCS which has the lowest spectral efficiency while fulfilling the condition shown in Equation (5.4).

5.4 Two-hop V2X communication over sidelink

As introduced in the last section, 5G should be designed to offer a solution to V2X communication with ultra-high reliability and low latency [MET13a, OBH+13]. Compared with the direct V2X communication, applying the LTE-Uu interface will introduce a relatively higher packet E2E latency since data packets need to be routed over network infrastructures [JWHS18b]. However, the required ultra-high reliability of V2X communication cannot always be fulfilled by a single-hop direct V2X communication [3GP17e]. For example, in order to assist a vehicle to obtain a good environment perception, a target V2X communication range up to 1000 m should be supported on the highway [3GP17e]. However, a single-hop direct V2X communication faces the challenge to achieve this large communication range since the Rxs far away from the Tx will statistically sense a bad signal quality. Thus, the packet transmission towards these Rxs has low reliability. To overcome this problem and increase the packet transmission range, a two-hop direct V2X communication over the sidelink is proposed by the author in this section. The packet transmission range refers to the maximal distance at which a V2X packet can be successfully delivered.

5.4.1 System model of two-hop direct V2X communication

To improve traffic safety and efficiency, the information of each vehicle and the road conditions, e.g. geographical location, mobility pattern and the sensed environment information, needs to be broadcasted to other traffic participants within a certain radius. This radius is often referred to as the V2X communication range whose value depends on the detailed service requirements [3GP17e]. As aforementioned, a collective perception of the environment (CPE) use case can contribute to a fully automated driving on a highway while a communication range of 1000 m is required [3GP17e]. The signal propagation loss can be quite high at such a large distance, and thus it cannot be guaranteed that the signal arrives at all relevant Rxs with a good quality for successful packet decoding.

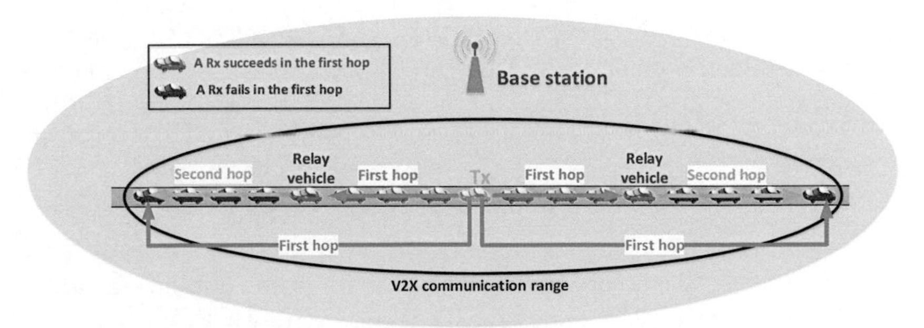

Figure 5.8: Two-hop direct V2X communication to enable local information exchange on highway [JDWS18]

Figure 5.8 illustrates the two-hop V2X communication proposed by the author to increase the packet transmission range and enhance the communication reliability. In this figure, a V2X Tx broadcasts its data packets to the nearby Rxs over sidelink communication. However, an Rx far away from the Tx may not successfully receive the packets from the single-hop transmission and it obtains a low packet reception ratio

(PRR). In this section, the direct transmission link between the packet generator, i.e. the V2X Tx, and an Rx is referred to as the first hop. In the proposed two-hop V2X communication scheme, after the transmission over the first hop, an Rx which successfully decoded the packets from the first hop can be configured as a relay to forward the packets to other Rxs over sidelink communication. Thus, the Rxs which failed in receiving the packets from the first hop can try to receive the packets from a nearby relay vehicle, and this link is referred to as the second hop in this work.
To enhance the efficiency of the proposed two-hop V2X communication scheme, there are some key design problems to cope with, as discussed below:

- How many vehicles should act as relays for a packet transmission? - The number of relay vehicles for each packet transmission needs to be reasonable. A large number of relay vehicles will introduce either additional interference or request more dedicated spectrum resources, depending on whether the different relay vehicles simultaneously use the same resource to transmit their packets or not.

- Which vehicle should act as a relay? - The relay vehicles for each packet transmission should be carefully chosen. If a relay vehicle is too close to the Tx, the coverage extension by applying the two-hop direct V2X communication over sidelink communication can be limited. In the other case, if a relay vehicle is too far away from the Tx, there is a high risk that it cannot successfully receive the packets from the first hop. As the PC5 interface for direct V2X communication is connection-less, there will be no feedback message from an Rx to the Tx, and the Tx is not aware whether a packet is received by a relay vehicle successfully or not.

- How to allocate radio resources among different transmission hops? - Since part of the overall system bandwidth is required to support the second hop V2X communication, the resource allocation between the first hop and the second one plays a critical role in system performance.

In order to take the above-mentioned aspects into account for applying the two-hop V2X communication, the author provides more insights in this section.

5.4.2 Context-aware radio resource management

From an efficiency perspective, the relay selection procedure can use context information to adapt itself to the concrete application scenario, e.g. geographical location of the nearby traffic participants, estimation of PRR, and environment. Therefore, a smart relay selection procedure contains two sub-tasks:

1. Selection of relay UE(s).

2. Collection of the useful context information.

As the first step, before a packet is transmitted over the first hop, a V2X Tx needs to locally decide which Rxs should act as relays. This relay selection procedure should try to serve the most relevant Rxs facing difficulty to successfully receive packets from the first hop. One way to achieve this target is to select a lot of vehicles as relays, because a small number of relays will increase the risk that some Rxs are not in the proximity of any relays. However, the number of relays cannot be arbitrarily large, since different relays will introduce mutual interference and consume more radio resources. In this section, since a large communication range is targeted in a highway scenario and the relevant Rxs locate either in the front side or in the rear side of the Tx, two vehicles will be selected as relays to transmit over the second hop, as shown in Figure 5.8. The following steps will be executed at a V2X Tx to select the two relays in this thesis:

1. A Tx needs to estimate the PRR of each nearby Rx.

2. The Tx searches for all Rxs with an estimated PRR higher than a predefined probability $Prob^{\text{threshold}}$ and considers them as relay candidates.

3. Among these relay candidates, the Rx with the largest distance to the Tx at each side, i.e. either front side or rear side, will be assigned as a relay.

In this work, a value of 99% is considered as the threshold value, i.e. $Prob^{\text{threshold}}$ in Step 2, to judge whether an Rx is suitable to act as a relay. Moreover, the relay selection procedure proposed by the author requires a V2X Tx to collect some context information. For example, the location of nearby Rxs should be predicted by the Tx to calculate their distances to the Tx. Since the real-time location and mobility pattern of the Tx are embedded in each V2X message, the V2X UE can receive and cache the information of its nearby traffic participants to predict their geographical location in the future. In addition, to estimate the PRRs of different Rxs, a mapping table from transmission distances to PRRs needs to be constructed locally at a V2X Tx. In this sense, the transmission of the CAMs can be used to obtain this table. As CAMs are a set of periodically transmitted messages from a V2X Tx, once a V2X UE receives a CAM message, it can predict the occurrence of the next CAM message. Thus, if the V2X UE successfully receives the next CAM message, it can record the reception status as successful. On the other hand, if the V2X UE does not receive the next CAM message, the reception status of that report is recorded as unsuccessful. In addition, since the location information of the Tx is embedded in the CAMs, a mapping table between the distances and the PRRs can be created. Please note that, since different MCSs exist in LTE, the mapping table should be created w.r.t. different MCSs. For clarification, if there are totally N_{MCS} optional MCSs and the PRRs should be collected within a communication range of 1000 m with a resolution of 5 m, the mapping table should have a size of $N_{\text{MCS}} \times 200$.

5.4.3 Transmission range maximization for two-hop V2X communication

5.4.3.1 Optimization problem

In order to increase the packet transmission range of the proposed two-hop direct V2X communication, the resource assigned to the two different hops should be adapted by the BS. Thus, a packet transmission range maximization problem can be constructed as

$$\begin{aligned} \textbf{maximize}_{(BW_1,BW_2)} \quad & D(BW_1) + D(BW_2), && (5.5)\\ \textbf{subject to} \quad & BW_1 + BW_2 \leq BW, && (5.6)\\ & BW_1 \geq 0, && (5.7)\\ & BW_2 \geq 0. && (5.8) \end{aligned}$$

The function $D(BW)$ in Equation (5.5) computes the radius inside which a packet can be successfully received by allocating a bandwidth of BW to this radio hop. Thus, if BW_1 corresponds to the spectral resources allocated to the first hop, $D(BW_1)$ stands for the packet transmission range for the first hop. It is assumed that there is a relay UE at the distance of $D(BW_1)$ from the original transmitter, in order to obtain the theoretical maximal transmission distance. Afterwards, the relay UEs located on the border of the transmission range will forward the received packets over the second hop, as proposed in the relay selection procedure in Section. 5.4.2. If a bandwidth of BW_2, which is orthogonal to the bandwidth for the first hop, is used for the second hop, $D(BW_2)$ is the packet transmission range for the second hop. Therefore, the optimization problem in Equation (5.5) tries to maximize the overall transmission distance by allocating bandwidth resource to the two different hops, i.e. BW_1 and BW_2. Moreover, since it is assumed that the two different hops operate on orthogonal frequency resources, the summation of the spectral resources of the two different hops should not exceed the overall available bandwidth, as shown in Equation (5.6). Moreover, the constraint functions in Equations (5.7) and (5.8) show that the allocated bandwidths cannot have negative values.
To derive the transmission distance function $D(BW)$, the general formula in Equation (5.9) is used to calculate the decimal value of pathloss,

$$PL = A_{\mathrm{p}} \cdot d^{\zeta}, \tag{5.9}$$

where A_{p} is constant in $\frac{1}{\mathrm{m}^{\zeta}}$ and ζ is the pathloss exponent. Moreover, as defined in Table 2.1, d denotes the distance between a Tx and its Rx in m. Please note that the values of A_{p} and ζ depend on the concrete environment model and the carrier frequency. By converting the above decimal value to its dB value, the pathloss models given in Table 2.2 can be obtained. With the help of this formula, the received signal power from the j-th Tx to the i-th Rx can be calculated as

$$P_{\mathrm{rx}}(i,j) = \frac{P_{\mathrm{tx}}}{A_{\mathrm{p}} \cdot d(i,j)^{\zeta}}. \tag{5.10}$$

$d(i,j)$ is the distance between the j-th Tx and the i-th Rx, and P_{tx} is the transmit power per RB which is assumed to be constant for all Txs in this section. Further, the SINR at the i-th Rx can be computed as

$$SINR(i,j) = \frac{P_{\text{tx}}/[A_{\text{p}} \cdot d(i,j)^{\varsigma}]}{\Sigma_{m \neq j}\{P_{\text{tx}}/[A_{\text{p}} \cdot d(i,m)^{\varsigma}]\} + \sigma_n^2}. \tag{5.11}$$

In this equation, the term "$m \neq j$" represents the other Txs transmitting their data over the same resource as the j-th Tx, and σ_n^2 is the noise power per RB. For simplification, $I(i,j)$ is the interference power in a case that the i-th Rx is trying to receive a packet from the j-th Tx, as

$$I(i,j) = \sum_{m \neq j} \{P_{\text{tx}}/[A_{\text{p}} \cdot d(i,m)^{\varsigma}]\}. \tag{5.12}$$

Therefore, Equation (5.11) can be simplified as

$$SINR(i,j) = \frac{P_{\text{tx}}/[A_{\text{p}} \cdot d(i,j)^{\varsigma}]}{I(i,j) + \sigma_n^2} \tag{5.13}$$

In order to successfully receive a packet from the j-th Tx, the SINR value experienced by the i-th Rx should be better than a threshold value. Theoretically, this SINR threshold value can be derived from Shannon's capacity equation as

$$C = BW \cdot \log_2(1 + SINR). \tag{5.14}$$

Equation (5.14) provides a theoretical limit on the capacity C over a bandwidth of BW with a given SINR. Please not that Equation (5.14) is only valid for Gaussian SINR and a Gaussian signal alphabet. In this section, the Shannon's capacity equation is used as an approximation. Thus, the SINR threshold value for successfully receiving a packet can be derived as

$$SINR(i,j) \geq 2^{(C/BW)} - 1. \tag{5.15}$$

Please note that the system capacity in Equation (5.15) should be calculated by Equation (5.3) based on the simulation parameters. By using Equation (5.13) to replace the left-hand side of Equation (5.15), the maximal distance at which a packet can be successfully received, i.e. the packet transmission range over a single-hop, can be calculated as

$$D(BW) = \frac{\sqrt[\varsigma]{P_{\text{tx}}/A_{\text{p}}}}{\sqrt[\varsigma]{[I(i,j) + \sigma_n^2] \cdot (2^{C/BW} - 1)}}. \tag{5.16}$$

As it can be seen from this equation, the transmission distance of a single-hop direct V2X communication is not only related with the allocated resource BW, but also with the interference power $I(i,j)$. The concrete interference power is directly impacted by the RRM scheme executed at the network side, and its details will be discussed later in Section 5.4.3.2. With Equation (5.16), the objective function given by Equation (5.5) can be re-written as

$$D(BW_1) + D(BW_2) = \frac{\sqrt[\varsigma]{P_{\text{tx}}/A_{\text{p}}}}{\sqrt[\varsigma]{[I(i,j) + \sigma_n^2] \cdot (2^{C/BW_1} - 1)}} + \frac{\sqrt[\varsigma]{P_{\text{tx}}/A_{\text{p}}}}{\sqrt[\varsigma]{[I(n,i) + \sigma_n^2] \cdot (2^{C/BW_2} - 1)}}. \tag{5.17}$$

Equation (5.17) represents the packet transmission distance of the proposed two-hop V2X communication. In this case, the j-th V2X UE sends its data packets towards other traffic participants, and the i-th V2X UE acts as a relay to forward these packets. Since the n-th Rx is far away from the j-th V2X UE, it fails in the reception from the first hop and, therefore, it receives the packets that are relayed by the i-th V2X UE.

5.4.3.2 Radio resource allocation between the two different hops

As mentioned previously, the resource allocation between the first hop and the second hop transmissions should be adapted to increase the packet transmission range. In Section 5.4.3.1, the author has constructed the objective function in Equation (5.17) and the corresponding constraint functions in Equations (5.6) to (5.8). Equation (5.17) reflects that the maximal transmission range is related to the sum of the interference power and the noise power. However, the interference power is not constant and it depends on the positions of the Txs transmitting over the same radio resource. As stated in Section 5.3, there are two transmission modes, i.e. the V2X sidelink transmission mode 3 and the V2X sidelink transmission mode 4, for sidelink V2X communication [3GP18e]. In the V2X sidelink transmission mode 3, the transmission resource is scheduled by the network and, therefore, the network can allocate the same radio resource to different Txs if the minimal distance between any two of them is larger than a certain threshold, e.g. the inter-site-distance. In mode 4, a V2X Tx needs to autonomously select a resource from resource

pools that are either configured by the network or pre-configured in the user device. In this case, the whole world can be divided into different zones and the same resource pool can be assigned to different geographical zones, if they are separated by a distance larger than a threshold. In both V2X sidelink transmission modes 3 and 4, the distance threshold value should be configured in a way that the interference power is low enough. In order to control the interference, the network needs to be aware of the pathloss model, and this information can be obtained from the V2X messages. Since reference signals are embedded in the packets transmitted over the PC5, a V2X UE can estimate the channels from its nearby V2X Txs by receiving their transmitted packets. Moreover, as location information is also contained in the V2X messages, the V2X UE can also calculate the transmission distances from the Txs and, therefore, compute the pathloss model. Afterwards, the V2X UE needs to report its formulated V2X pathloss model to the network, and the network can utilize this information to compute the distance threshold value so that the interference is under good control.
In this section, the distance threshold value is set to be the inter-site distance. The interference power is at least 3 dB lower than the noise power. In this case, the objective function in Equation (5.17) can be simplified to

$$\begin{aligned} D(BW_1) + D(BW_2) \approx & \frac{\sqrt[\alpha]{P_{\text{tx}}/A_{\text{p}}}}{\sqrt[\alpha]{\sigma_n^2 \cdot (2^{C/BW_1} - 1)}} \\ & + \frac{\sqrt[\alpha]{P_{\text{tx}}/A_{\text{p}}}}{\sqrt[\alpha]{\sigma_n^2 \cdot (2^{C/BW_2} - 1)}}. \end{aligned} \tag{5.18}$$

With the constraint functions in Equation (5.6) to (5.8), the optimization problem can be solved by setting its derivative equal to zero and thus the objective function in Equation (5.18) achieves its maximal value if $BW_1 = BW_2$.

5.4.4 Simulation parameters and numerical results

In order to evaluate the proposed two-hop V2X communication, a system-level simulator is implemented to simulate a highway scenario with 3 lanes and a length of 20 km. As mentioned in Section 2.3, the Hata channel model is used to calculate the pathloss for sidelink communication in this scenario. Moreover, the BSs are deployed with an ISD of 6 km alongside the highway for providing CP connection to V2X UEs. The V2X sidelink transmission mode 3 is used for data transmission in the UP, and it operates on the 2 GHz carrier. In addition, two inter-vehicle distance (IVD) options, i.e. 10 m and 15 m, are used, and each vehicle has a length of 5 m. Please note that the number of vehicles under the coverage of a cell, i.e. N_v in Equation (5.3), should be calculated by taking account of the number of lanes, the IVD and the ISD. For instance, if an IVD of 10 m is assumed, there are $\frac{3\times6000}{10} = 1800$ vehicles in a cell and the required system capacity C in Equation (5.3) can be calculated:

$$
\begin{aligned}
C &= 212 \text{ byte/packet} \times 8 \text{ bit/byte} \times 10 \text{ packet/s} \times 1800 \\
&= 30\,528 \text{ kbps}
\end{aligned}
\tag{5.19}
$$

Besides, the 1×2 antenna configuration, i.e. receiver diversity, is exploited for the direct V2X communication over the sidelink. Last but not least, it is assumed that packets should be delivered to all traffic participants located within a 1000 m radius of the Tx.

Table 5.3: System performances of different resource allocation schemes (inter-vehicle-distance is 10 m) [JDWS18]

BW_1	BW_2	PRR (IVD = 10 m)
5 MHz	0 MHz	39.67%
6 MHz	0 MHz	45.74%
8 MHz	0 MHz	54.26%
10 MHz	0 MHz	56%
5 MHz	5 MHz	77.72%
6 MHz	4 MHz	77.31%
8 MHz	2 MHz	70.90%

In the following, the performance of different schemes will be shown. At first, the performance of the single-hop direct V2X communication will be compared with that of the developed two-hop scheme in Table 5.3 and Table 5.4, in terms of packet reception ratio. The difference between these two tables is the different inter-vehicle distance, which introduces a different V2X UE density. Moreover, different bandwidth resources are allocated to the first hop and the second hop transmission in Table 5.3 and Table 5.4, in order to validate the approach proposed for maximizing the packet transmission distance. Following that, Table 5.5 compares the

Table 5.4: System performances of different resource allocation schemes (inter-vehicle-distance is 15 m) [JDWS18]

BW_1	BW_2	PRR (IVD = 15 m)
5 MHz	0 MHz	54.26%
6 MHz	0 MHz	57.54%
8 MHz	0 MHz	67.02%
10 MHz	0 MHz	71.28%
5 MHz	5 MHz	100%
6 MHz	4 MHz	99.82%
8 MHz	2 MHz	84.93%

performances of two different schemes for allocating resource the relay transmitters in the second hop, i.e. two relay transmitters are assigned with the same resource, and the two relay transmitters are assigned with orthogonal resources.
The PRRs of different resource allocation schemes are given in Table 5.3, where the IVD is set to 10 m. In this table, BW_1 and BW_2 represent the number of spectral resources allocated to the first hop and the second hop, keeping in mind that there are two relay vehicles for the second hop transmission and they simultaneously use the same resource for their second hop transmission. The schemes with "$BW_2 = 0$ MHz" apply the single-hop V2X communication, since no resource is allocated to the second hop. It can be seen that the single-hop direct V2X communication has low PRRs, since the target communication range of 1000 m cannot be achieved. In addition, different resource allocation schemes for the two-hop V2X communication are investigated by the author to validate the analysis developed in Section 5.4.3.2. An efficient approach to increase the transmission range of the single-hop communication is to increase the system bandwidth. The reason is that a larger system bandwidth enables a Tx to use an MCS with better robustness. Alternatively, the proposed two-hop direct V2X communication can utilize the spectral resource in an efficient manner. For instance, a PRR of 56% can be achieved by a single-hop direct V2X communication with 10 MHz system bandwidth. In comparison, if the two-hop direct V2X communication is applied with the same amount of bandwidth, the transmission range can be efficiently improved. In this approach, the 10 MHz bandwidth is divided into two non-overlapping subsets. In Table 5.3, the best performance is achieved with an equal resource allocation between the

two different hops, i.e. 5 MHz for the first hop and another 5 MHz for the second hop, and the PRR is increased from 56% in the single-hop case to 77.72%. This observation validates the mathematical analysis provided in Section 5.4.3.2. Moreover, it is also worth noticing that the PRR, i.e. 77.72%, in the two-hop transmission scheme with an equal resource allocation between different hops are slightly lower than two times of the PRR, i.e. 39.67%, in the single-hop transmission scheme with 5 MHz bandwidth. The logic here is that the strongest interference for the second hop transmission is from the other relay vehicle transmitting the same packet, and the distance between two relay vehicles transmitting the same packet has a value lower than two times the communication range, i.e. 2000 m in the considered scenario. However, the interference for the first hop transmission is from Txs served by other BSs, and thus the distance between two nearby interfering Txs for the first hop has statistically the same average value as the inter-site distance, i.e. 6000 m in the considered scenario. Therefore, a stronger interference can be experienced for the second hop than for the first hop, and the transmission range of the second hop is thus smaller than that of the first hop, though the two different hops are assigned the same amount of resources. In Table 5.4, the IVD is increased from 10 m to 15 m and, thus, there are fewer vehicles deployed in the simulator. The lower system capacity C allows using an MCS with better robustness for V2X communication. Therefore, the PRRs in Table 5.4 are higher than those in Table 5.3. Moreover, the PRR of using single-hop V2X communication with 10 MHz bandwidth is 71.28%, and it can be increased to 100% by the proposed two-hop scheme with an equal resource allocation to the two different hops. As a compromise, the packets successfully received from the second hop will experience a higher E2E latency than the packets received from the first hop. However, since the Rxs failing in the first hop receptions are statistically far away from the Tx, their reaction time can be relaxed, and a certain performance deterioration w.r.t. E2E latency is tolerable.

In Table 5.5, the author inspects the two resource allocation schemes for the second hop. In both schemes, 10 MHz is the allocation to the first hop. Additionally, in the first scheme, another 10 MHz bandwidth is dedicated to the second hop where the two relays transmit the same packet over the same resource. In this sense, the strongest interference for the second hop comes from the other relay transmitting the same packet. In the second scheme, the 10 MHz bandwidth is also allocated to the second hop, but the two relays transmitting the same packet will use different sub-bands, i.e. 5 MHz for one relay and another 5 MHz for the other one. Therefore, there is no mutual interference between the two relays transmitting the same packet. Compared with the first scheme, though the interference power of the second hop is now lower, there is less resource allocated to each packet transmission over the second hop and an MCS with a worse robustness needs to be used. Thus, the PRR in the second scheme has a lower value than the first resource allocation scheme. In other words, though a mutual interference will be introduced for the case where the two second hop links reuse the same time-and-frequency resource, its system performance in the considered scenario is still better than the other scheme, where dedicated and orthogonal resources are assigned to the different second hop transmitters.

Table 5.5: System performance comparison of different resource allocation schemes for the second hop transmission (inter-vehicle-distance is 10 m) [JDWS18]

BW_1	BW_2	PRR (IVD = 10 m)
10 MHz	10 MHz for both relays	99.98%
10 MHz	5 MHz for one relay and another 5 MHz for the other relay	95.90%

5.5 Applying multi-RAT to enhance the reliability of V2X communication

Previously, the author has introduced the two-hop V2X communication to improve the communication reliability on a highway. In this scenario, the BSs are deployed alongside the highway with an inter-site distance up to several km. Thus, the direct V2X communication via the PC5 has its advantages compared with a cellular V2X communication via the LTE-Uu. For example, a V2X Tx on the cell border experiences a low efficiency to transmit packets to its serving BS due to a high pathloss, while a direct V2X communication does not face this problem. Therefore, a good approach to configure the network in this scenario is to

facilitate the UP data transmission via the PC5, while the LTE-Uu interface is exploited for CP functions. However, in other scenarios, e.g. in an urban scenario, the ISD can be several hundred meters which is comparable to the target V2X communication range. Thus, it is worth applying the LTE-Uu for the UP transmissions.
As mentioned before, the ultra-high reliability requirement cannot be always fulfilled by a single-RAT. For instance, in a conditional automated driving use case, ultra-high reliability with a communication range of 278 m is required in a city to avoid any application-layer message retransmissions [3GP17e]. Later in Section 5.5.4, where the system performance will be provided, the challenges to fulfill the ultra-high reliability requirement by a single-RAT can be observed. Therefore, a multi-RAT scheme to improve the communication reliability is proposed by the author in Figure 5.9, where a V2X data packet is transmitted both via the LTE-Uu and PC5 interfaces. Compared with the single-RAT schemes shown in Figures 5.2 and 5.6, an Rx in the proposed multi-RAT scheme receives the same data packet from two different air interfaces, i.e. the LTE-Uu and the PC5.
It is worth noticing that additional resources are required by the proposed multi-RAT transmission scheme, compared with the single-RAT schemes. However, it is considered as the most straightforward approach in system design to improve communication reliability by consuming additional resources. For instance, in order to improve the robustness of mission-critical communications, the multi-connectivity technology applies a transmit diversity by using additional resources [MET17a].

5.5.1 Multi-RAT V2X communication

5.5.1.1 Independent transmissions over the LTE-Uu and the PC5

The V2X Tx needs to reside in the RRC_CONNECTED state in order to send an SR message to the BS for requiring the uplink transmission resource. Thus, although the sidelink transmission over the PC5 can support V2X Txs in both the RRC_CONNECTED state, i.e. by sidelink transmission mode 3, and the RRC_IDLE state, i.e. by sidelink transmission mode 4, the V2X Tx in the multi-RAT scheme needs to enter the RRC_CONNECTED state before obtaining the transmission resource for the uplink. An independent multi-RAT transmission via both the

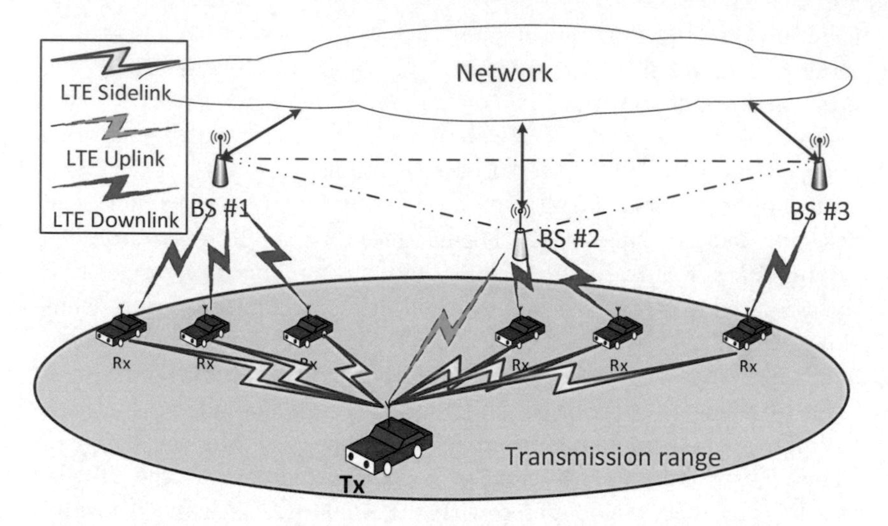

Figure 5.9: V2X communication over both the LTE-Uu and the PC5 in UP [JWHS18a]

LTE-Uu and the PC5 can be enabled by extending the current 3GPP approach [3GP18f]. The signaling diagram to support the independent V2X transmissions over sidelink transmission mode 3 and the LTE-Uu is proposed and developed by the author as shown in Figure 5.10. The signaling diagram is illustrated in the following:

1. The V2X Tx decides to apply the multi-RAT transmission scheme, if ultra-high communication reliability is required by the service. Afterwards, the V2X Tx in the RRC_IDLE state performs the connection establishment procedure to transit to the RRC_CONNECTED state.
2. The V2X communication can be carried out via the LTE-Uu and the PC5 independently, with more details listed below:
   2.A Regarding the V2X communication over the LTE-Uu interface, the Tx sends an SR message to the BS.
   2.B Regarding the V2X sidelink communication, as the sidelink transmission mode 3 is applied, a Sidelink UE information message is sent from the Tx to the BS for sidelink transmission resource acquisition [3GP18f].
3. The BS schedules transmission resources for the V2X communica-

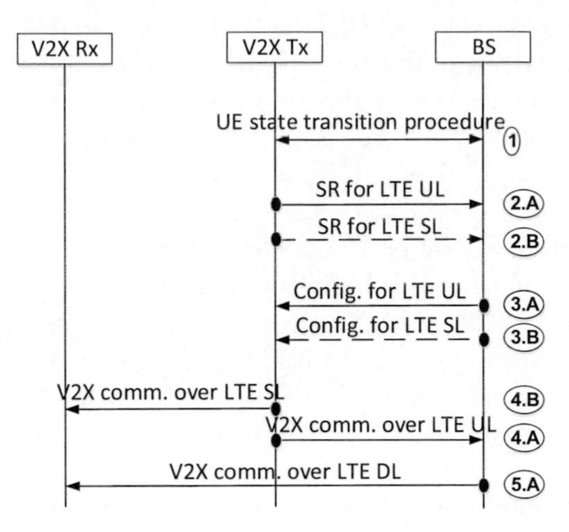

Figure 5.10: Signaling diagram for independent multi-RAT transmission over the LTE-Uu and the PC5 [JWHS18a]

tion over the LTE-Uu interface and over the sidelink communication. Then

3.A the BS sends a message containing the resource configuration information for the cellular V2X communication over the LTE-Uu back to the Tx.

3.B the BS sends a message containing the resource configuration information for the direct V2X communication over the PC5 back to the Tx.

4. With the resource configuration information,

   4.A the Tx can send its V2X message in the uplink to the BS.

   4.B the Tx multicasts its V2X message over the PC5 interface.

5. Once the BS successfully receives the packet in the uplink,

   5.A the packet will be further forwarded to other relevant BSs and then sent in the downlink to the respective Rxs as stated in Section 5.2.1.

In order to separate the steps used for the communications via the LTE-Uu and the PC5 clearly in Figure 5.10, letters "A" and "B" are attached to the end of the sequential indices, correspondingly. The sequential indices ended with letter "A" refer to the communication steps related with the transmission via the LTE-Uu, while the letter "B" corresponds to the transmission via the PC5. Please also note that these two letters are applied in the rest of this section with the same intention. Moreover, since the communications in this multi-RAT transmission scheme take place independently, there is no sequential relationship between any two steps from the two different RATs. For instance, it is possible that the V2X Tx transmits the SR message to require a resource for the LTE uplink, i.e. Step 2.A, after the transmission over the PC5, i.e. Step 4.B, is accomplished. In addition, the above signaling scheme takes place when a data packet arrives at the buffer of the Tx while there is no granted resource for its transmission. This can refer to the cases where the V2X Tx initiates a new data transfer or the SPS configuration information has expired.

5.5.1.2 Coordinated transmissions over the LTE-Uu and the PC5

In order to obtain the resource configuration for the two different air interfaces, the independent multi-RAT transmission scheme introduced before has the drawback that two scheduling request procedures should be conducted. Thus, a large signaling overhead for the V2X communication will be introduced if the dynamic scheduling approach is applied or the resource configured by the SPS has expired. For instance, the event-driven type of V2X traffic corresponds to the emergency messages if certain hazardous situations are detected by a traffic participant. In this case, since a V2X Tx needs to dynamically request the transmission resource, the independent multi-RAT scheme foresees a large signaling overhead. In order to reduce the signaling effort, the author proposes and develops a coordinated multi-RAT transmission scheme in Figure 5.11 that enables coordinated packet transmissions via the LTE-Uu and the PC5. The signaling diagram is detailed below:

1. The V2X Tx first needs to establish a connection with the network and enter the RRC_CONNECTED state.
2. An SR message will be sent to the network for transmission resource acquisition.
3. The BS sends a message containing the resource configuration information for both the cellular V2X communication and direct V2X communication back to the Tx.
4. With the resource configuration information,

   4.A the Tx can send its V2X message in the uplink to the BS.

4.B the Tx multicasts its V2X message over the PC5 interface.

5. Once the BS successfully receives the packet in the uplink,

   5.A the packet will be further forwarded to other relevant BSs and then sent in the downlink to the respective Rxs as stated in Section 5.2.1.

Since Step 2 is relevant for the decision of whether a single-RAT or multi-RAT transmission should be applied, there are two options as listed below.

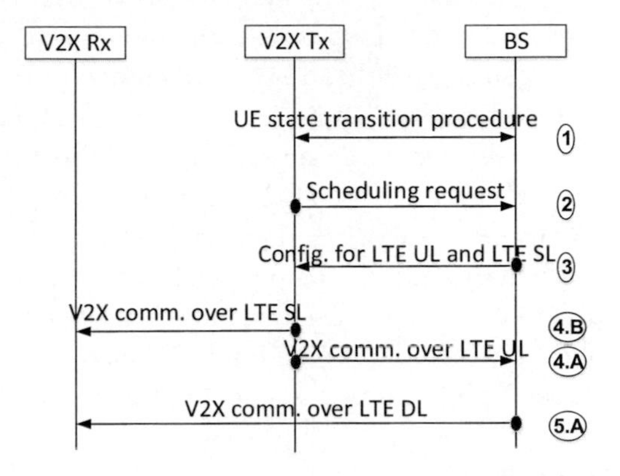

Figure 5.11: Signaling diagram to support the coordinated multi-RAT transmission [JWHS18a]

1. In the first option, the traffic profile information, e.g. traffic type and QoS requirements, will be embedded in the SR message for resource scheduling. Thus, based on this information, the BS analyses and checks if a multi-RAT transmission should be exploited for better communication reliability. If so, the BS sends the configuration information back to the V2X Tx where the transmission resource information for both the LTE-Uu and the PC5 is carried.

2. The second option is to perform a local analysis at the V2X Tx to derive the most appropriate transmission scheme. If a multi-RAT transmission scheme is selected, the V2X Tx sends an SR message to request the transmission resources for both the LTE-Uu and the PC5 interfaces. In this message, certain context information, e.g. the geographical position of the Tx and the cellular pathloss value,

can be carried in order for the BS to perform an efficient resource scheduling for the different interfaces. Finally, the configuration information will be sent back from the BS to the V2X Tx.

Compared with the independent multi-RAT transmission approach, the coordinated approach has the advantages that only one SR message needs to be sent to the BS and the BS can jointly schedule the transmission resources for both the LTE-Uu and PC5 interfaces. Therefore, this scheme can efficiently reduce the signaling overhead while exhibiting the capability to coordinate the transmissions via different interfaces. Please note that the low signaling overhead is an advantage in the CP, and the performance of the different schemes, i.e. independent transmission and the coordinated transmission, in the UP will be the same since they both transmit a packet via the LTE-Uu and PC5 interfaces. In Section 5.5.4, the performance of these two proposed multi-RAT schemes in the UP will be demonstrated. To be specific, in the downlink of the LTE-Uu interface, both the unicast and multicast transmission modes are inspected and they are correspondingly labeled as "direct V2X + unicast" and "direct V2X + multicast" in Figures 5.19, 5.20, and 5.21.

5.5.2 Hybrid uplink

As mentioned before, compared with a single-RAT scheme, a multi-RAT scheme consumes a larger spectrum resource due to the duplicate transmissions via different interfaces. However, since the spectrum resource of an operator is limited, an efficient spectral usage is essential for V2X communication.

5.5.2.1 Hybrid uplink procedure: exploiting the transmission over the PC5 interface

As introduced in Section 5.3, the sidelink transmission enables a local data exchange over a single radio hop. Compared with the Rxs far away from the Tx, the Rxs located near the Tx will statistically obtain better reliability from sidelink communication due to better radio conditions. In addition, compared with the cellular V2X communication via the LTE-Uu, the lower E2E latency achieved from the sidelink communication helps the V2X Rxs to take actions in a more timely manner. Since the Rxs within a short range of the Tx are more sensitive to packet E2E

latency, they demand a low latency and a highly reliable V2X communication to avoid accidents. Therefore, the sidelink communication is of high importance for them. On the other hand, up to 3GPP Release 15, the uplink transmission from a traffic participant to a BS only takes place via the Uu interface [3GP16a]. However, there is always a possibility that the signal transmitted over the sidelink can arrive at the BS with certain strength. This offers an opportunity for the network to receive data from a traffic participant without using the uplink spectrum. A prerequisite in this case is that a BS should be equipped with the capability to receive data packets over the PC5.
It is worth noticing that a V2X UE is proposed by 3GPP to have the ability for transmitting over both the LTE-Uu and the PC5 interfaces [3GP16a]. In addition, it is also proposed in 3GPP that an RSU is able to receive V2X messages via different interfaces, e.g. PC5 or LTE-Uu, depending on the implementation option [3GP18a]. Therefore, to equip a BS with the transmission capability via both interfaces is not a big implementation issue.

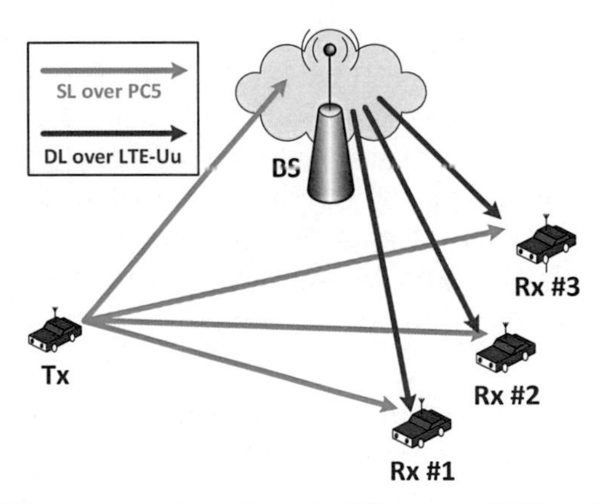

Figure 5.12: V2X communication where the BS receives V2X packets via the PC5 interface [JWHS18a]

In Figure 5.12, the multi-RAT scheme is proposed by the author where a BS receives data packets from the PC5 interface. For simplicity, only one BS is plotted in this figure. In this scheme, the packets transmitted via the PC5 interface not only arrive at the traffic participants, but

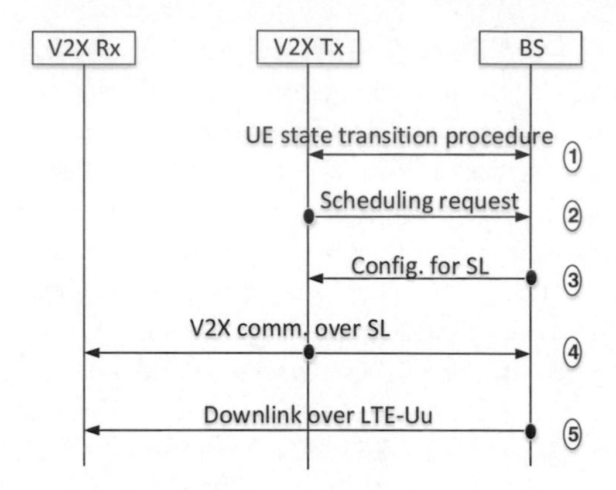

Figure 5.13: Signaling diagram to enable multi-RAT V2X communication where the BS receives data packets from the PC5 [JWHS18a]

also at the BS. Therefore, if the strength of the received signal is high enough, it can be successfully received by the BS equipped with the sidelink communication capability. Afterwards, the received packets will be transmitted through the network and delivered to other relevant BSs for the downlink transmissions over the LTE-Uu as introduced in Section 5.2.1.3. Moreover, the corresponding signaling diagram to support this scheme is provided in Figure 5.13. Since the BS in this scheme will only configure the Tx with the transmission resource for sidelink communication, the V2X data packets will be received by the BS from the sidelink communication, and the BS acts as one of the ordinary Rxs. The signaling diagram in Figure 5.13 are provided with more details in the following:

1. The V2X Tx first needs to establish a connection with the network and enter the RRC_CONNECTED state.

2. An SR message will be sent to the network for transmission resource acquisition.

3. The BS sends a message containing the resource configuration information for the direct V2X communication back to the Tx.

4. With the resource configuration information, the Tx multicasts its V2X message over the PC5 interface.

5. Once the BS successfully receives the packet via the sidelink communication, the packet will be further forwarded to other relevant BSs and then sent in the downlink to the respective Rxs by using the LTE-Uu interface.

Compared with the scheme shown in Figure 5.11, this scheme has the advantage that no dedicated uplink resource is required and data packets are transmitted to the network over the sidelink resource. This procedure of using the PC5 interface for transmission towards the BS is different from the conventional uplink procedure where the LTE-Uu interface is applied. Thus, the conventional uplink procedure needs to be extended. In this section, the term "hybrid uplink" represents the concept that both the LTE-Uu and the PC5 interfaces can be considered as the options for a BS to receive the transmission from a UE. It is worth noticing that, as the sidelink communication corresponds to a P2MP transmission up to 3GPP Release 15, there is no ACK/NACK message fed back from an Rx to the Tx, and the transmission reliability from a Tx to the BS cannot be guaranteed.
Please also note that the above proposed hybrid uplink transmission over the PC5 is labeled as "PC5 for uplink" in Figures 5.22 and 5.23, where its performance will be provided.

5.5.2.2 Hybrid uplink: multi-RAT to improve uplink reliability

If a data packet is transmitted through the network infrastructure, the V2X communication is composed of two radio transmission hops, i.e. uplink and downlink. Since the downlink transmission takes place sequentially after a successful uplink transmission, a packet failed in the uplink transmission will introduce a serious performance degradation in the V2X communication. For instance, as shown in Figure 5.12, a packet failed in its uplink transmission from the vehicle Tx to the BS cannot be further transmitted to the relevant Rxs in the downlink. Thus, a PRR of 0% can be foreseen for this packet transmission, regardless of the performance in the downlink. In the other case where the BS in the scenario of Figure 5.12 successfully receives the uplink packet while one downlink transmission fails, the PRR can still reach 66.67%. From this example it can be seen that a packet transmission failure in the uplink causes a more serious performance degradation than a failure in the downlink

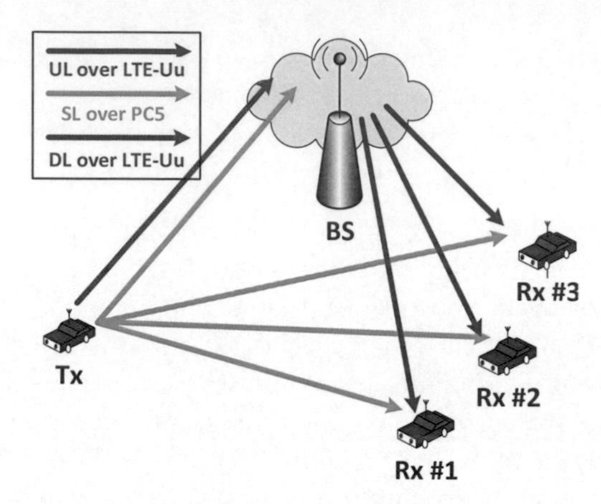

Figure 5.14: V2X communication where the network receives V2X packets via both the PC5 and the LTE-Uu uplink interfaces [JWHS18a]

w.r.t. the PRR, and thus extra effort should be spent to enhance the reliability of the uplink transmission.
In Section 5.5.2.1 the hybrid uplink scheme has been introduced by the author to exploit both the LTE uplink and the sidelink for the transmission from a Tx to its serving BS. Therefore, a multi-RAT transmission scheme proposed by the author as shown in Figure 5.14 can be applied for an enhanced uplink and better communication reliability. In this scheme, data packets will be transmitted from a Tx to the BS over both the LTE-Uu and the PC5, which is the difference from the scheme shown in Figure 5.12. Moreover, since the sidelink transmission over the PC5 in this scheme refers to a multicast transmission where the BS acts as one of the ordinary Rxs, a dedicated resource is only required for the uplink transmission over the LTE-Uu.

5.5.2.3 Coordination between different RATs in hybrid uplink

In order to efficiently support the hybrid uplink transmission over both the LTE-Uu and the PC5 interfaces in the scheme of Figure 5.14, two different signaling schemes are proposed by the author.

5.5.2.3.1 Independent hybrid uplink transmission scheme In this approach, both the LTE uplink and the sidelink in Figure 5.14 will be

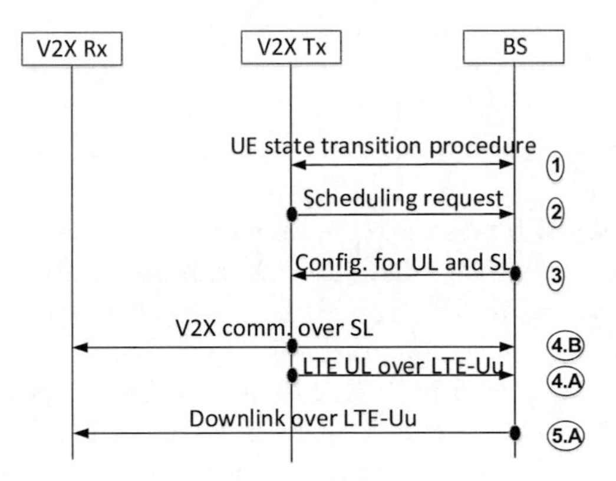

Figure 5.15: Signaling diagram to support the independent hybrid uplink transmission scheme [JWHS18a]

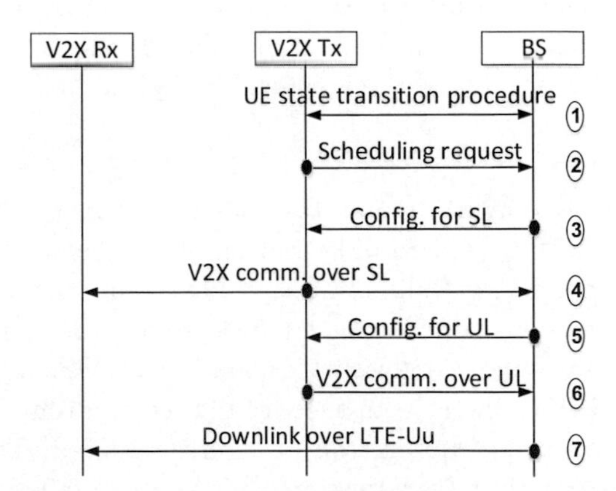

Figure 5.16: Signaling diagram to support the sequential hybrid uplink transmission scheme [JWHS18a]

independently configured for packets transmitted to the BS. Correspondingly, the signaling diagram is proposed by the author as given in Figure 5.15, and it is detailed in the following:

1. The V2X Tx first needs to establish a connection with the network and enter the RRC_CONNECTED state.
2. An SR message will be sent to the network for transmission resource acquisition.
3. The BS sends a message containing the resource configuration information for both the cellular V2X communication and direct V2X communication back to the Tx.
4. With the resource configuration information,
   4.A the Tx can send its V2X message in the uplink to the BS.
   4.B the Tx multicasts its V2X message over the PC5 interface, and the BS acts as an ordinary Rx for sidelink communication.
5. Once the BS successfully receives the packet either from the uplink or from the sidelink,
   5.A the packet will be further forwarded to other relevant BSs and then sent in the downlink to the respective Rxs.

As it can be seen, both transmissions over the LTE uplink and the sidelink can take place immediately after obtaining the resource configuration information from the BS, and the BS will try to receive data packets from both the uplink and sidelink interfaces. In this thesis, the author names this scheme as the independent hybrid uplink transmission scheme.

5.5.2.3.2 Sequential hybrid uplink transmission scheme One advantage of the independent hybrid uplink transmission scheme is to configure the resource for the LTE uplink and the sidelink in a flexible way, since the two transmissions are decoupled from each other. On the other hand, as the LTE uplink is always exploited to provide a diversity gain regardless of whether the BS can successfully receive from the sidelink or not, a large resource for uplink transmission is required. To counter this problem, the LTE uplink transmission can be sequentially triggered after a data packet is not successfully received by the BS from the sidelink. The author proposes the signaling diagram in Figure 5.16 to support this sequential hybrid uplink transmission scheme. The signaling diagram is detailed below:

1. The V2X Tx first needs to establish a connection with the network and enter the RRC_CONNECTED state.
2. An SR message will be sent to the network for transmission resource acquisition.
3. The BS sends a message containing the resource configuration information for the direct V2X communication back to the Tx.
4. With the resource configuration information, the Tx multicasts its V2X message over the PC5 interface.
5. Together with other Rxs located in the proximity of the Tx, the BS will try to receive the packets from the sidelink. In case the BS successfully receives the packet from the sidelink, the LTE uplink transmission will not be triggered, and the BS proceeds with the downlink transmission as the signaling diagram will be the same as the one shown in Figure 5.13. In the other case where the BS fails in receiving a packet from the sidelink, it sends another message to the Tx to trigger a transmission over the LTE-Uu uplink. The transmission resource over the LTE-Uu is indicated by this message.
6. With the configuration information for the transmission over the LTE-Uu, the Tx transmits the packet to the BS in the LTE-Uu uplink.
7. Once the BS successfully receives the packet via the sidelink communication, the packet will be further forwarded to other relevant BSs and then sent in the downlink to the respective Rxs by using the LTE-Uu interface.

Since the LTE-Uu uplink corresponds to a P2P transmission, its MCS can be adjusted based on the estimated channel quality. However, as the MCSs in the LTE are designed to operate within a certain SINR range [JF12], there is a chance that an LTE uplink channel experiences an SINR value that is even worse than the lower bound of the operation range w.r.t. SINR and, thus, none of the MCSs can provide a good radio link robustness. As a solution in this case, the Tx needs to select the most robust MCS, i.e. the one with the lowest modulation and coding rate, and perform the blind retransmissions of the same packet over

uplink. At the receiver side, i.e. the BS, the MRC procedure will be performed to enhance the uplink reliability.
To support the transmission redundancy introduced by the blind retransmission scheme, an additional frequency resource is required. However, in the sequential hybrid uplink transmission scheme, since the LTE-Uu uplink transmission is only triggered for the packets that are not successfully received from the sidelink, there are fewer packets transmitted over the LTE-Uu uplink compared with the independent hybrid uplink transmission scheme. Namely, more resources will be available for each packet transmission over the LTE-Uu uplink in the sequential hybrid uplink transmission scheme, and that provides good support for the blind retransmission scheme. On the other hand, since the LTE-Uu uplink transmission is executed sequentially after the sidelink transmission, the latency to receive a packet from the LTE-Uu uplink in the sequential transmission scheme is large. Compared with the independent hybrid uplink transmission scheme, an additional delay component for the LTE-Uu uplink transmission is deduced, and its value corresponds to a round trip time of the sidelink communication. In detail, this delay component can be decomposed into a packet transmission time, a time duration of two processing procedures, and a signaling message transmission time. The packet transmission duration is related to many parameters, e.g. the packet size, the efficiency of the applied MCS and the number of copies retransmitted blindly. Additionally, a duration of 3 ms is reserved for a processing procedure in the LTE, and the transmission duration of a signaling message can be 1 ms. Therefore, the proposed sequential transmission scheme introduces a minimal additional delay of 7 ms plus the packet transmission duration for the packets received from the LTE-Uu uplink, compared with the independent hybrid uplink transmission scheme.
Please note, the independent hybrid uplink transmission scheme and the sequential hybrid uplink transmission scheme proposed in this subsection are implemented in the system-level simulator. And their performances w.r.t. packet E2E latency and packet reception ratio will be evaluated and labeled as "PC5 + LTE-Uu for uplink" and "PC5 + enhanced LTE-Uu for uplink" later in Figures 5.22 and 5.23, correspondingly.

5.5.3 Fast MBSFN area mapping

Previously, the blind retransmission is used in the sequential hybrid uplink transmission scheme to enhance the reliability in the uplink. In the other transmission direction, i.e. the downlink, the Rxs located on the cell border will experience bad channel conditions due to the low received power of the desired signal and the strong superposed interference. Thus, it is critical to improve the downlink communication reliability for the cell-border Rxs.

In the LTE network, the multicast-broadcast single-frequency network (MBSFN) is a transmission technology enabling multiple BSs to synchronously multicast the same content. In this approach, the area covered by the synchronized BSs is referred to as an MBSFN area. In order to avoid the inter-symbol interference (ISI) in an MBSFN area, the transmissions from the different BSs need to be well synchronized so that the maximal delay spread at an Rx should be within the cyclic prefix duration of one OFDM symbol. In this manner, the received signals from different BSs can be constructively superposed at the Rx $\#i$ and its effective SINR can be calculated as

$$SINR_i = \frac{\sum_{j \in MBSFN}(|h_{i,j}|^2 P_j)}{\sum_{k \notin MBSFN}(|h_{i,k}|^2 P_k) + \sigma_n^2}. \tag{5.20}$$

The term "$j \in MBSFN$" represents the synchronized BSs belonging to the MBSFN area while "$k \notin MBSFN$" stands for the interfering BSs. From this equation, it can be seen that the MBSFN technology empowers an Rx to utilize the signals received from different synchronized BSs instead of considering them as distortion. Thus, with a collection of more cells to synchronously multicast the same data packets, a larger improvement w.r.t. the SINR can be expected. However, an MBSFN area cannot be arbitrarily large since that will bring certain drawbacks. For example, since different propagation delays are introduced by the transmissions from different BSs, a large MBSFN area also causes a large delay spread at the Rx. In this case, in order to avoid ISI, a longer cyclic prefix duration is necessary which will decrease the transmission efficiency. Therefore, the size of an MBSFN area has clearly an impact on the system performance. In this thesis, an MBSFN area is restricted to the coverage area of three cells.

In order to carry out the MBSFN procedure, a BS needs to forward its received data packets to the V2X server which will process the data and further distribute them to the relevant traffic participants [3GP15c]. Therefore, the V2X server can derive the MBSFN area by analyzing the context information, e.g. user position and cell ID, carried in each data packet. After that, the network forwards the packet to the relevant BS(s). To reduce the packet E2E latency, two fast MBSFN area mapping procedures enabling a local data exchange among different BSs are developed by the author as shown in Figures 5.17 and 5.18. In these figures, a cell layout of three sectors per BS is used and the cells indicated with the same color form an MBSFN area. Furthermore, the V2X Tx is located in the coverage area of cell #1 and served by the BS #1. The two proposals for MBSFN area mapping are detailed below.

- **Fixed MBSFN area mapping based on the serving cell index**: Upon the successful packet reception from the uplink, the packet is synchronously transmitted by the BS #1 over its three cells, i.e. cells #1, #2 and #3 in Figure 5.17. In this approach, an MBSFN area refers to the coverage area of the cells operated by the serving BS of the Tx.

- **Dynamic MBSFN area mapping based on channel estimation**: By performing channel estimation on the CSRS, the V2X Tx can be aware of the propagation conditions from the different cells. Thus, a V2X Tx indicates the indices of the three cells with the lowest propagation losses when it transmits its data packets to the BS, i.e. the cells #1, #4 and #5 in Figure 5.18. These cells create the MBSFN area for packets generated by the V2X Tx. Once a packet of this V2X Tx is successfully received by the BS #1, the packet will be directly routed from BS #1 to BS #2 and multicasted in cells #1, #4 and #5 in a synchronous manner. In this scheme, the MBSFN area refers to the coverage area of the cells providing the best channel conditions to the V2X Tx.

In the first proposal, i.e. the fixed MBSFN area mapping, the function to coordinate resource usage among different BSs can be deployed at the edge of the RAN, e.g. at the BS. Thus, this proposal can simplify the network architecture to support MBSFN without demanding a frequently-changed MBSFN area. The MBSFN area is only changed if the V2X Tx enters the coverage of another BS. In comparison, the second proposal,

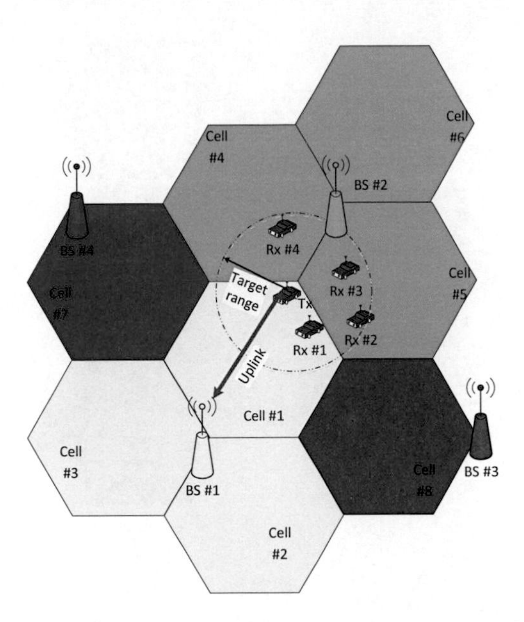

Figure 5.17: Fixed MBSFN area mapping based on the serving cell index [JWHS18a]

i.e. the dynamic MBSFN area mapping, poses a higher complexity on the network. At first, context information, i.e. the propagation losses from different BSs, is required to derive the MBSFN area. In addition to that, a packet needs to be routed dynamically from one BS to another, and a synchronized transmission over the time-and-frequency domain has to be achieved among different BSs belonging to the same MBSFN area. However, as shown in Figure 5.18, the synchronized transmissions from cells #1, #4 and #5 can contribute to higher SINR values for the V2X Rxs located on the borders of these cells. To be noticed, these V2X Rxs are located most close to the V2X Tx and, therefore, ultra-high communication reliability should be guaranteed. In this case, the higher SINR values can enhance the transmission robustness for these Rxs.
Please also note, in Section 5.5.4 the author applies both the above proposed MBSFN area mapping approaches along with the direct V2X communication over the sidelink to inspect the performance of the corresponding multi-RAT schemes. They are labeled as "direct V2X + fixed MBSFN" and "direct V2X + dynamic MBSFN" in Figures 5.19, 5.20,

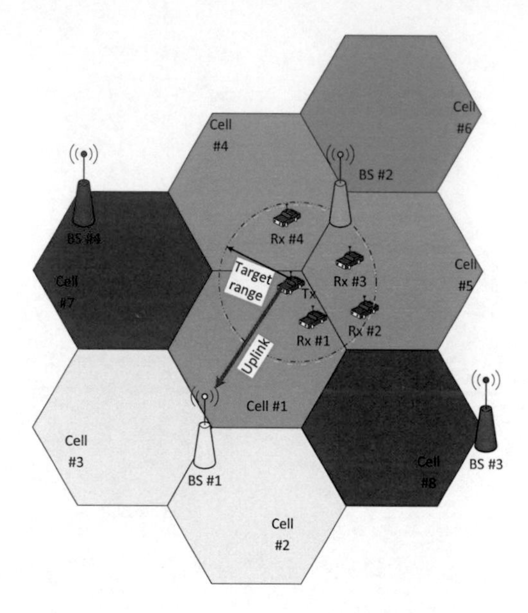

Figure 5.18: Dynamic MBSFN area mapping based on channel estimation [JWHS18a]

and 5.21, respectively.

5.5.4 Simulation parameters and numerical results

In order to evaluate the proposed technology in an urban scenario, a simulator is implemented where macro BSs are deployed in the Madrid-grid model shown in Figure 2.1. The LTE-Uu interface operates on a carrier frequency of 2 GHz with a total bandwidth of 20 MHz, i.e. 10 MHz/10 MHz for uplink/downlink, dedicated to the V2X communication. Additionally, the transmission on PC5 is over 5.9 GHz with a bandwidth of 10 MHz. Both LOS and NLOS propagation detailed in Section 2.3.3.2 are modeled to derive the pathloss values. To be more specific, the macro outdoor-to-outdoor propagation model in Section 2.3.3.2.1 is used to calculate the pathloss for cellular V2X communication over the LTE-Uu interface, while the direct V2X propagation model in urban scenario stated in Section 2.3.3.2.5 is applied for direct V2X communication over the PC5 interface.

Furthermore, the sidelink transmission mode 3 is used for direct V2X communication over the PC5, and, therefore, the resource for sidelink transmission is centrally scheduled by the network. Specifically, an SPS [3GP18c] algorithm is applied where the overall resource is evenly allocated to different V2X Txs. For instance, if there are 10 Txs with a packet transmission periodicity of 10 Hz in the system, then each Tx periodically gets a resource of 10 ms to transmit one packet. In case a packet is not successfully received in the allocated resource, it is considered being dropped, and its E2E latency is considered to be infinite. Moreover, since the focus in this thesis is on the RAN, the message transition latency among different BSs is not inspected in detail. Both the CP and UP functionalities of the BM-SC and MBMS-GW are assumed to locate at the edge of the RAN [3GP16a], and a latency value of 1 ms is assumed for the message transition among BSs. Additionally, as mentioned in Sections 5.2.1 and 5.3, both the sidelink communication and the LTE MBMS correspond to a multicast transmission mode. Thus, the two MCSs with a spectral efficiency of 0.6016 bit/Hz and 0.887 bit/Hz are exploited for the sidelink and the LTE MBMS, respectively.

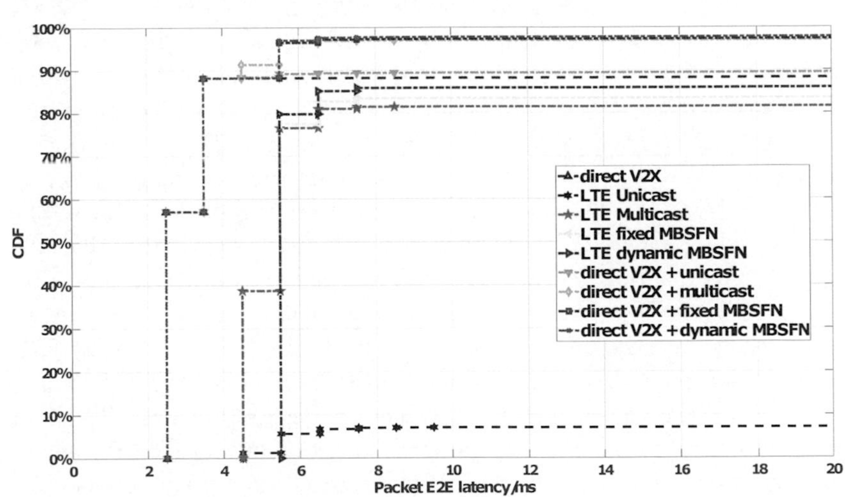

Figure 5.19: CDF of packet E2E latency (Target communication range = 200 m, and 1000 UEs per $(\text{km})^2$) [JWHS18a]

The CDF of the packet E2E latency is plotted in Figure 5.19 where the target V2X communication range is set to be 200 m and a vehicle density of 1000 vehicles per $(\text{km})^2$ is assumed. In this figure, the performance of the direct V2X communication over the PC5 and the performance of the LTE-Uu interface by using both the unicast and multicast transmission modes in the downlink are provided. Please note that the CDF curves do not converge to 100% since there are packets failed in their transmission and, therefore, the PRR can also be reflected in this figure. As can be seen, the LTE unicast mode has the worst performance due to its large resource requirement in the downlink. Comparing to that, the LTE multicast in the downlink is more resource-efficient in the considered V2X scenario and, therefore, it has a better performance w.r.t. the packet E2E latency and the PRR. In addition, the two different MBSFN area mapping approaches stated in Section 5.5.3 outperform the LTE multicast scheme due to the synchronized transmissions from different BSs and, therefore, the V2X Rxs on the cell border experience better SINR values with the MBSFN technology. More precisely, the PRR can be improved from 82% in the LTE multicast scheme to 86% in the LTE dynamic MBSFN area mapping scheme. Moreover, taking account of the radio condition experienced by the V2X Tx, the dynamic MBSFN area mapping approach provides a better robustness and, therefore, contributes to a higher PRR than the fixed MBSFN area mapping approach. As mentioned before, the V2X communication refers to a local information exchange procedure and, therefore, the direct V2X communication over the PC5 can provide a good performance within a moderate communication range. This point is also illustrated in Figure 5.19, as the PRR for the direct V2X communication over the PC5 is higher than the V2X communication schemes through the network infrastructure. Besides, since the data packets are not transmitted through the network infrastructure in the direct V2X communication scheme, its packet E2E latency is shorter than that of other schemes utilizing the LTE-Uu and it can fulfill the packet E2E latency requirement of 5 ms [MET13a]. In order to improve reliability, the performances of four multi-RAT schemes are also given. In the first multi-RAT scheme, i.e. labeled as "direct V2X + unicast", the V2X packets are transmitted over both the PC5 and the LTE-Uu unicast interfaces, and the PRR is better than in the case if the packets travel through a single-RAT. However, as the LTE-Uu unicast provides a comparably low PRR, the improvement from the multi-RAT is very slight compared with the performance of the direct V2X com-

munication. In the other three multi-RAT schemes, both sidelink and LTE-Uu multicast schemes are exploited, i.e. downlink multicast without SFN, multicast with fixed MBSFN area mapping, and multicast with dynamic MBSFN area mapping. As can be seen from the curves, the PRR can be improved from 88% in the single-RAT scheme to 97% in the multi-RAT scheme. It can also be noticed that the performance difference between the "direct V2X + fixed MBSFN" multi-RAT scheme and the "direct V2X + dynamic MBSFN" multi-RAT scheme is only 1% w.r.t. the PRR, and it is much smaller than in the case if the direct V2X communication is not used, i.e. a performance difference of 4%. This is due to the fact that many vulnerable cell-border UEs can successfully receive the data packets from the PC5.
In Figure 5.20, the V2X communication range is increased to 300 m and

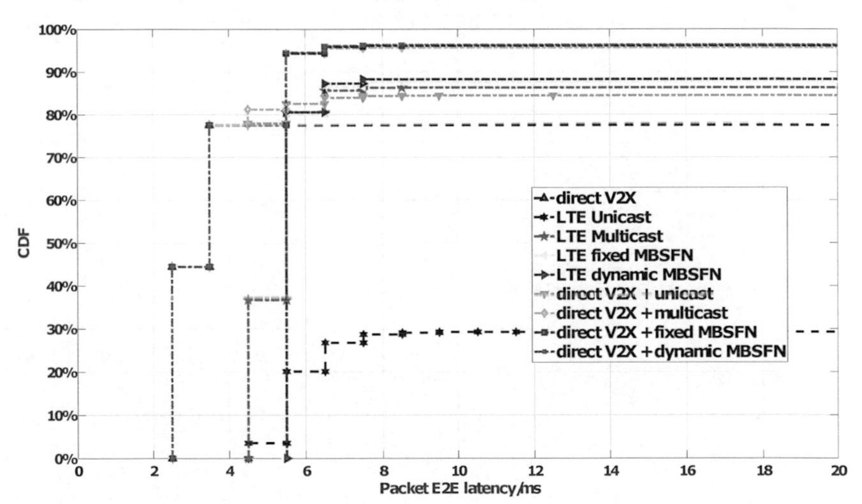

Figure 5.20: CDF of packet E2E latency (Target communication range = 300 m, and 500 UEs per $(\mathrm{km})^2$) [JWHS18a]

the vehicle density is decreased to 500 vehicles per $(\mathrm{km})^2$. As can be seen, the performance of the V2X communication through the network infrastructure, i.e. the LTE unicast, LTE multicast, LTE fixed MBSFN and LTE dynamic MBSFN schemes, are better than the ones shown in Figure 5.19, since a lower data volume contributes to a larger allocated resource for each packet transmission. However, the PRR of the direct

V2X communication is worse than that shown in Figure 5.19. This is due to the larger communication range and due to the fact that the V2X Rxs located far from the Tx experience worse radio conditions. Thus, the LTE multicast and the LTE MBSFN schemes outperform the direct V2X communication w.r.t. the PRR in this specific case. Additionally, both the LTE multicast and the direct V2X communication have a PRR worse than 86%. Again, by applying the multi-RAT scheme, i.e. direct V2X + LTE dynamic MBSFN area mapping, the PRR can be efficiently improved to be over 96%.

In order to observe the performances of different schemes w.r.t. dif-

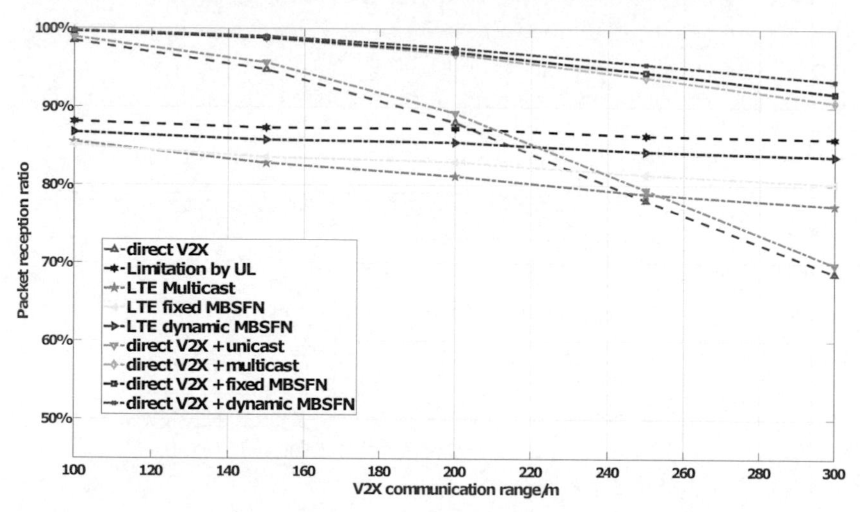

Figure 5.21: PRRs of different multi-RAT technologies w.r.t. the different communication ranges (1000 UEs per $(\text{km})^2$) [JWHS18a]

ferent communication ranges, the PRRs of the different schemes are plotted in Figure 5.21, i.e. from 100 m to 300 m with a step-width of 50 m. The vehicle density is set to be 1000 vehicles per $(\text{km})^2$. It can be observed that the performance of the direct V2X communication is significantly influenced by the communication range, since the signal propagation distance has an impact on the radio condition of the direct V2X communication. In comparison, the LTE-Uu multicast scheme is less sensitive to the communication range, as the signal propagation distance between a V2X UE and its serving BS is independent of the

communication range. By comparing the different multi-RAT schemes, it can also be seen that the multi-RAT scheme of using unicast in the downlink has clearly a worse performance than the other schemes due to its low efficiency in the downlink. In addition, the other schemes where V2X communication goes through the network infrastructure, i.e. LTE multicast, fixed MBSFN area mapping, and dynamic MBSFN area mapping, have more outstanding performance difference with an increased communication range and, therefore, the same tendency can be observed from the corresponding multi-RAT schemes. Last but not least, the performance limitation posed by the LTE uplink, i.e. labeled as "Limitation by UL", is also shown in this figure by assuming that all the packets can be successfully received in the downlink. As shown in this figure, the introduced dynamic MBSFN area mapping scheme approaches the uplink limitation quite well and, therefore, its performance is mainly limited by the transmission failures occurring in the uplink.
In Figure 5.22, the performances of different hybrid uplink transmission

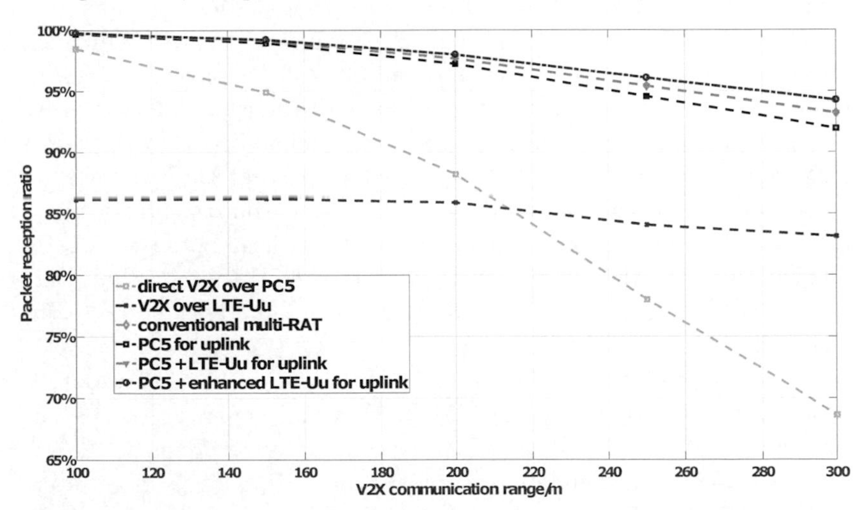

Figure 5.22: PRRs of different hybrid uplink technologies w.r.t. the different communication ranges (1000 UEs per $(km)^2$ and MBSFN is applied in the LTE-Uu downlink) [JWHS18a]

schemes are provided where the downlink transmission is realized by the dynamic MBSFN area mapping approach. For instance, the curve labeled "PC5 for uplink" shows the performance of the multi-RAT scheme where the BS receives the packets of a V2X Tx from the sidelink carrier, i.e. as stated in Section 5.5.2.1. And the curve labeled "PC5 + LTE-Uu for uplink" represents the multi-RAT transmission scheme where both the PC5 and the LTE-Uu are applied in the hybrid uplink transmission, i.e. the scheme shown in Section 5.5.2.2. In another case, the curve labeled "PC5 + enhanced LTE-Uu for uplink" shows the sequential hybrid uplink transmission scheme introduced in Section 5.5.2.3.2 where the transmission over the LTE-Uu uplink is only sequentially triggered after the BS does not successfully receive a V2X packet from the sidelink carrier. At the same time, the performances of the two single-RAT, i.e. direct V2X communication over the PC5 interface and V2X communication over the LTE-Uu interface with MBSFN in the downlink, are also provided as comparisons. Besides, the performance of the independent multi-RAT transmission scheme without using the hybrid uplink transmission is also plotted, i.e. labeled as "conventional multi-RAT". Comparing the curve where the BS only receives the V2X packets from the PC5 with the curves of the single-RAT schemes, shows a significant improvement of the PRR. Therefore, additionally exploiting the LTE-Uu downlink resource and equipping BSs with the capability to receive a sidelink transmission contributes to better reliability than the direct V2X communication. However, the performance of the two schemes where the packets are received at the BS either from the LTE-Uu, i.e. the curve shown by "conventional multi-RAT", or from both the LTE-Uu and the PC5, i.e. the curve shown by "PC5 + LTE-Uu for uplink", are consistent and, therefore, these two curves overlap. This is due to the fact that the additional transmission over the PC5 can hardly provide any contribution if the uplink transmission over the LTE-Uu interface is unsuccessful. In this simulation, the PC5 interface operates on the carrier frequency 5.9 GHz, which is higher than the carrier frequency 2 GHz of the LTE-Uu and, therefore, the signal propagation loss is more severe on the sidelink. Therefore, if a packet fails in its transmission over the LTE-Uu uplink, its transmission to the BS over the PC5 will very likely experience a failure too. Comparing to that, since the sequential hybrid uplink transmission scheme labeled as "PC5 + enhanced LTE-Uu for uplink" conditionally triggers the uplink transmission over the LTE-Uu, a better usage of the spectral resource can be achieved and, therefore, the

PRR can be improved from 93% in the conventional multi-RAT scheme to 94%, if the communication range is up to 300 m.
Besides the PRRs shown in Figure 5.22, the packet E2E latency of the

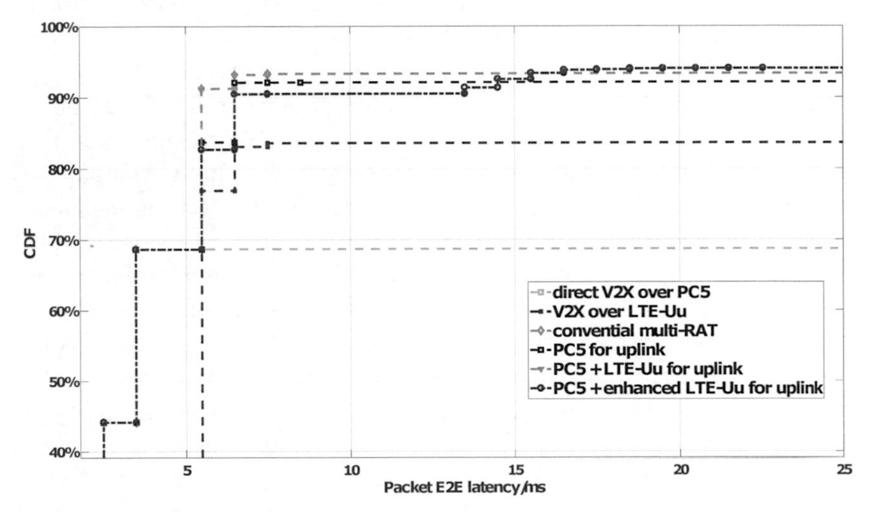

Figure 5.23: CDF of packet E2E latency w.r.t. different hybrid uplink technologies (Target communication range = 300 m, 1000 UEs per $(\text{km})^2$, and MBSFN is applied in the LTE-Uu downlink) [JWHS18a]

different hybrid uplink technologies is also of interest. In this sense, the CDFs of their packet E2E latency are provided in Figure 5.23, where the target communication range is set to 300 m and a vehicle density of 1000 vehicles per $(\text{km})^2$ is used. By looking at the low E2E latency scale in Figure 5.23, e.g. below 10 ms, it can be seen that the PRR of the sequential hybrid uplink transmission, i.e. the curve labeled "PC5 + enhanced LTE-Uu for uplink", within the low E2E latency scale is lower than the curves of the other hybrid schemes. This is due to the fact that the packets which are successfully received by the BS from the LTE-Uu experience a large E2E latency value. As mentioned in Section 5.5.2.3.2, in the sequential hybrid uplink transmission scheme the packet transmission in the uplink over the LTE-Uu only takes place if the packet has not been successfully received by the BS over the PC5 interface. Thus, the packets which are successfully received by the BS from the LTE-Uu interface experience a minimal additional delay of 8 ms, i.e. 7 ms + the minimal packet transmission duration of 1 ms, compared with the other hybrid uplink technologies. In addition, please note that the PRR of the sequential hybrid uplink transmission scheme in the low E2E latency scale is even lower than that of the case where the BS receives the packets only from the sidelink carrier, i.e. the curve labeled "PC5 for uplink". This is true because there are more packets successfully received in the uplink from the sequential hybrid uplink transmission scheme and thus more packets will be transmitted in the downlink. With more packets to be transmitted in the downlink, less resources will be allocated per packet transmission and the packets successfully received by the BS from the sidelink carrier will experience a performance degradation in the downlink. However, as mentioned before, the sequential hybrid uplink transmission scheme can contribute to a high PRR but with a large E2E latency. This point can be reflected by inspecting the relatively high packet E2E latency scale in Figure 5.23. In general, the performance shown in Figure 5.22 validates the ability of the developed sequential hybrid uplink transmission scheme to improve the communication reliability by efficiently utilizing the uplink resource.

5.6 Summary

At the beginning of this chapter, the author has introduced the different cellular technologies, i.e. the LTE-Uu and the PC5, to enable V2X communication. In particular, the author described both the unicast and multicast transmission modes of the LTE-Uu interface with the focus on their applications in V2X communication. Besides, as the PC5 interface is standardized in 3GPP to enable the ProSe between two nearby devices, the author has also inspected its application in the direct V2X communication.

In order to increase the packet transmission range of traffic-related data packets, the author has proposed a two-hop V2X communication. As some V2X communication applications require a large communication range which a single-hop direct V2X communication cannot achieve, an increased packet transmission range by the proposed two-hop transmission technology can contribute to a higher PRR than the single-hop transmission scheme. To exploit the two-hop direct V2X communication in an efficient manner, the author has provided the detailed analysis of

the resource allocation scheme. In addition, context information has been collected and taken into account to select proper relays. In order to evaluate the proposed technology, a system-level simulator has been implemented to inspect the performance of different V2X communication schemes in a highway scenario. The simulation results have shown the performance improvement by applying the two-hop direct V2X communication over the sidelink with the same amount of spectral resource as for single-hop direct V2X communication. For instance, if the IVD is set to 10 m, the PRR is improved from 56% to 77.72%. In the other case where the IVD is 15 m, the two-hop direct V2X communication can increase the PRR from 71.28% to 100%. In addition, the results have also shown that the performance of the two-hop direct V2X communication can be optimized by adapting the resource allocation for the different hops.
Moreover, in order to provide a better reliability for V2X communication in an urban scenario, a multi-RAT scheme has been proposed where packets are transmitted through both the LTE-Uu and the PC5 interfaces. Correspondingly, the author has designed different signaling schemes to compromise between flexibility and signaling efficiency. In addition, in order to protect the packet transmission from a V2X Tx to its serving BS, the conventional uplink transmission over the LTE-Uu interface has been extended to a hybrid uplink transmission technology where the BS can flexibly receive data packets through either the LTE-Uu or the PC5 or both of them simultaneously. In order to show the performances of the different technologies, the author has also implemented a system-level simulator and provided the simulation results. It can be seen that the performances of the different technologies are related to the concrete application scenarios. For instance, the direct V2X communication over the PC5 provides a low E2E latency and good reliability to the Rxs within a moderate communication range, i.e. up to 250 m in the considered urban scenario. However, its performance sensitively degrades with an increased communication range. In comparison, the V2X communication through the network infrastructure can utilize the LTE-Uu interface to multicast the data packets in the downlink and it can contribute to a better reliability within a large communication range, i.e. from 250 m in the considered urban scenario. In order to meet the ultra-high reliability requirements in some V2X scenarios, the

multi-RAT transmission over both the PC5 and the LTE-Uu can be applied. If a BS is able to receive the packets from the PC5 interface, the proposed multi-RAT scheme has the ability to facilitate the transmission through the network infrastructure without using the uplink resources and it can still contribute to a large reliability improvement, compared with a single-RAT scheme. For instance, if a V2X communication range of 300 m is required, the single-RAT scheme provides a PRR of 68% by direct V2X communication and a PRR of 83% by cellular V2X communication. Compared to that, the proposed multi-RAT scheme can increase the PRR to 92% without using the uplink resources. Meanwhile, when uplink resources are available in the network, the hybrid uplink scheme can also be applied to further improve the communication reliability in an efficient manner, i.e. the PRR with a V2X communication range of 300 m is increased from 92% to 94% in the considered scenario. Last but not least, the communication reliability in the LTE-Uu downlink has been enhanced by the single-frequency network technology where the radio propagation information is taken into account to coordinate the transmissions from different BSs. For example, if a V2X communication of 200 m is needed, the cellular V2X communication with a multicast downlink mode via the LTE-Uu air interface provides a PRR of 82% in the considered scenario. As a comparison, this value can be increased to 86% if the dynamic MBSFN approach is applied.

Chapter 6
Conclusion

It is expected that the direct communication mode among nearby devices in 5G will not only improve the system performance of current human-type services, but also provide performance enhancement for emerging new types of services, such as IoT and automated driving services. This thesis addresses concepts on how to improve the performance of different 5G service types by D2D communication, i.e. eMBB, mMTC and V2X, and evaluates them quantitatively.

6.1 Summary

The aforementioned service types for 5G have divergent QoS requirements. For example, compared with legacy 4G system, 5G targets at offering an enhanced QoS for MBB services with higher spectral efficiency and system capacity. In the mMTC use case, battery life and network coverage are the two important factors influencing the market penetration of 5G systems, while the data volume is not the main concern. Regarding V2X communication, ultra-high reliability and low latency should be guaranteed by 5G. As the author has inspected the applicability of D2D communication in different use cases, this thesis can be correspondingly summarized in the following.

Network-controlled sidelink communications to offload network traffic

1. In this part, the author has proposed RRM algorithms with low computational complexity for allocating resources to underlaying D2D links. The low computational complexity provides a low delay to configure D2D links with a valid response. The proposed RRM algorithms are able to provide improved system performance w.r.t. the number of established D2D links and system capacity.

2. Moreover, since the mobile users in reality have different QoS requirements regarding their data rate and because they might be assigned different priority levels, the author has also proposed a

smart RRM algorithm that is able to minimize the amount of required resources to support the service requirements of all links. In its decision making, the smart RRM algorithm is able to take additional context information such as service requirements and user priorities into consideration. The proposed RRM algorithm determines whether a link operates in a dedicated mode or a reuse mode. If a link operates in the reuse mode, the algorithm further decides which other link shall share the same RB.

3. In order to support the proposed RRM algorithms, the author has designed novel signaling schemes to assist D2D communication in both single-cell and multi-cell scenarios. The signaling schemes have the ability to collect the context information required by the RRM algorithms with a reasonable signaling effort.

4. As a proof-of-concept, simulations are carried out. The numerical results have shown a better system performance by the proposed RRM algorithms compared with baseline schemes in terms of the number of established D2D links, i.e. performance gain of 68%, overall system capacity, i.e. performance gain of 20%, and user satisfaction ratio, i.e. users with higher priority are served with better QoE.

Network-controlled sidelink communication in the mMTC use case

1. In this part of the thesis, mMTC services have been enhanced by a network-controlled sidelink communication scheme. In this concept, it has been proposed that some sensors can be assigned by the network to act as relays to forward packets of remote sensors which experience bad cellular channel conditions.

2. In order to ensure a good efficiency for D2D communication, different clustering algorithms have been applied to group mMTC devices, and only an intra-group D2D communication is allowed. Moreover, the mode configuration of different devices has been designed by analyzing the user-specific context information.

3. Three critical signaling procedures to support the proposed sidelink communication have been described, i.e. initial attachment of sen-

sors, update of the TM, and the uplink report by sidelink communication procedures. Meanwhile, the proposed signaling schemes enable the collection of the required context information without a heavy signaling overhead.

4. The simulation-generated numerical results have shown that the network-controlled sidelink communication improves both the network coverage, i.e. from 86% to 98.5%, and the amount of mMTC devices fulfilling the battery life requirement, i.e. from 80.5% to 95% at the same time.

Network-controlled V2X communication

1. In this work, this author has inspected the cellular V2X communication, i.e. via the LTE-Uu, and the direct V2X communication, i.e. via the PC5, standardized by 3GPP for enabling V2X communication. In particular, the author has described the current approach to apply the LTE-Uu air interface with both the downlink unicast and multicast transmission modes in V2X communication. Moreover, the direct V2X communication via the PC5 air interface has also been detailed in this thesis.

2. As some V2X communication applications require a large communication range that a single-hop direct V2X communication cannot achieve, a two-hop V2X communication over the sidelink has been developed. In this approach, relay vehicles can be assigned to forward packets to the remote vehicles over the second hop. Context information has been collected and taken into account to select the proper relays. Moreover, in order to improve the efficiency and achieve a maximal packet transmission range, the resource allocation to the different radio hops has been optimized.

3. In addition, to provide better reliability for V2X communication, a multi-RAT scheme has been proposed where packets are transmitted through both the LTE-Uu and the PC5 interfaces. Different signaling schemes have been designed to compromise between configuration flexibility and signaling efficiency. Besides, the author has extended the conventional uplink transmission to a hybrid uplink transmission technology where the BS is able to receive data packets via either the LTE-Uu or the PC5 or both of them simultaneously.

The developed approaches can improve the V2X communication reliability in terms of packet reception ratio.

4. Last but not least, a system-level simulator has been implemented to provide the simulation results that validate the performance gains of the technical proposals compared with the state-of-the-art technologies. For instance, if the requirement of the communication range is up to 300 m, the proposed multi-RAT transmission scheme can achieve a packet reception ratio of 94% in the considered scenario, comparing to the packet reception ratio of 69% by using PC5 and 83% by using the LTE-Uu interface.

6.2 Future work

The technical proposals in this thesis have demonstrated the advantages of applying a direct communication mode between two nearby user devices in 5G. However, certain topics relevant to D2D and V2X communications are still to be solved as a continuation of this thesis.

- In this thesis, the author has developed the resource allocation schemes exploiting context information, e.g. location information of user devices, and channel gain. However, a perfect knowledge of location information cannot be assumed, and its accuracy depends on the applied positioning technology, e.g. the Global Positioning System (GPS). Moreover, the channel estimation at the receivers cannot be always considered as perfect in a real system. Thus, the impact of using inaccurate context information should be inspected.
- In Chapter 4, sidelink communication has been applied to facilitate the direct communication between a remote sensor device and a relay sensor device. It was assumed that the direct links and the cellular links operate with two orthogonal sets of resources and thus there is no mutual interference between a direct link and a cellular link. However, since the remote sensors experience a high pathloss to their serving BSs, the signal power arriving at the BSs from remote sensors is very low. This provides the opportunity for the network to assign the same time-and-frequency resource to both a direct link transmission and a cellular link transmission, as long as

they are far away from each other. One challenge here is to design efficient signaling schemes for keeping the energy consumption of the CP functions low enough, as it is highly relevant for calculating the battery life of sensor devices.

- As shown in Chapter 5, different RATs, e.g. cellular V2X and direct V2X, have different performances w.r.t. latency and reliability for V2X communication. On the other hand, different V2X applications have also different service requirements. Thus, an efficient RA selection for each UE and each application is critical for the system performance and it needs to be developed in future work. An optimal RAT selection scheme should take account of context information, such as CSI and geographical location of UEs. Moreover, since the UEs in V2X communication normally have high mobility, the network topology varies dynamically with the time, and the RAT selection procedure needs to take place frequently. Thus, a signaling scheme with low effort to configure the proper RATs for the UEs should be developed.

Moreover, alongside with the research work of this thesis, it has been realized by the author that a technically sound and efficient D2D communication in the next generation mobile system cannot be achieved by just adding a number of functions in the system design. Instead, a crafted architecture considering all aspects from the physical layer to the RRC layer should be developed. For instance, the following topics should be inspected in the future.

- Mobility control for D2D communication: The D2D mobility control should be natively supported by 5G [SJD+18]. The traditional mobility control solutions target at a reliable and uninterrupted connection between a UE and a network access point, as they are the two ends of the communication link. However, the mobility control in D2D communication is more sophisticated, as more than one UE is involved. For instance, when a D2D transmitter enters the coverage of a new cell while the D2D receiver remains in the old cell, how should the network coordinate the two different cells and perform the D2D handover procedure?

- As mentioned before, the mMTC use case requires a massive amount of connections to connect sensors and machines. This can lead to an instant burst of RA requests and introduce a serious RA collision problem [HS17]. In order to mitigate this problem, a local RA procedure can be applied where an mMTC device can be configured as an aggregation point for its nearby devices to forward their RA requests to the BS. Since UE power efficiency is another key factor in the mMTC use case, a power-efficient signaling scheme needs to be designed to support this group-based RA procedure.

- A coexistence of multiple service types is foreseen in 5G. This scenario does not only apply for cellular links but also for D2D links. For example, a passenger in a vehicle might require MBB services over a direct communication mode with the passengers in other vehicles for entertainment, e.g. video chatting or gaming, and meanwhile a V2X communication also takes place to facilitate the automated driving. Thus, it is essential to enable a simultaneous usage of multiple service types in 5G. The key challenge here is how to efficiently manage the radio resource to serve the different service types [HJS18].

References[1]

[3GP10] 3GPP. Evolved Universal Terrestrial Radio Access (E-UTRA); Further advancements for E-UTRA physical layer aspects. Technical Report TR 36.814, Third Generation Partnership Project (3GPP), March 2010.

[3GP13a] 3GPP. Channel models for D2D performance evaluation. Technical Report R1-132030, Third Generation Partnership Project (3GPP), May 2013.

[3GP13b] 3GPP. Discussion on UE-UE Channel Model for D2D Studies. Technical Report R1-130092, Third Generation Partnership Project (3GPP), January 2013.

[3GP13c] 3GPP. Study on provision of low-cost Machine-Type Communications (MTC) User Equipments (UEs) based on LTE (Release 12). Technical Report TR 36.888, Third Generation Partnership Project (3GPP), June 2013.

[3GP14] 3GPP. Study on LTE Device to Device Proximity Services-Radio Aspects. Technical Report TR 36.843, Third Generation Partnership Project (3GPP), March 2014.

[3GP15a] 3GPP. Cellular system support for ultra-low complexity and low throughput Internet of Things (CIoT) (Release 13). Technical Report TR 45.820, Third Generation Partnership Project (3GPP), November 2015.

[3GP15b] 3GPP. New WI Proposal: D2D based MTC. Technical Report RP-151948, Third Generation Partnership Project (3GPP), December 2015.

[3GP15c] 3GPP. TP for 36.885. Technical Report R1-155411, Third Generation Partnership Project (3GPP), October 2015.

[3GP16a] 3GPP. Study on architecture enhancements for LTE support of V2X services. Technical Report TR. 23.785, Third Generation Partnership Project (3GPP), September 2016.

[3GP16b] 3GPP. Study on new services and markets technology enablers. Technical Report TR 22.891, Third Generation Partnership Project (3GPP), September 2016.

[3GP16c] 3GPP. Study on Scenarios and Requirements for Next Generation Access Technologies (Release 14). Technical Report TR 38.913, Third Generation Partnership Project (3GPP), October 2016.

[3GP16d] 3GPP. User Equipment (UE) radio access capabilities (Release 13). Technical Report TR 36.306, Third Generation Partnership Project (3GPP), December 2016.

[3GP17a] 3GPP. Feasibility study for evolved Universal Terrestrial Radio Access (UTRA) and Universal Terrestrial Radio Access Network (UTRAN). Technical Report TR. 25.912, Third Generation Partnership Project (3GPP), March 2017.

---

1 The 3GPP documents can be found in 3GPP portal: https://portal.3gpp.org/

[3GP17b] 3GPP. Group Communication System Enablers for LTE (GCSE_LTE); Stage 2. Technical Report TS. 23.468, Third Generation Partnership Project (3GPP), December 2017.

[3GP17c] 3GPP. Multimedia Broadcast/Multicast Service (MBMS); Architecture and functional description. Technical Report TS. 23.246, Third Generation Partnership Project (3GPP), December 2017.

[3GP17d] 3GPP. Proximity-based services (ProSe); Stage 2. Technical Report TS 23.303, Third Generation Partnership Project (3GPP), June 2017.

[3GP17e] 3GPP. Study on enhancement of 3GPP support for 5G V2X services. Technical Report TR. 22.886, Third Generation Partnership Project (3GPP), March 2017.

[3GP18a] 3GPP. Architecture enhancements for V2X services. Technical Report TS 23.285, Third Generation Partnership Project (3GPP), March 2018.

[3GP18b] 3GPP. Evolved Universal Terrestrial Radio Access (E-UTRA) and Evolved Universal Terrestrial Radio Access Network (E-UTRAN); Overall description; Stage 2. Technical Report TS. 36.300, Third Generation Partnership Project (3GPP), January 2018.

[3GP18c] 3GPP. Evolved Universal Terrestrial Radio Access (E-UTRA); Medium Access Control (MAC) protocol specification. Technical Report TS 36.321, Third Generation Partnership Project (3GPP), July 2018.

[3GP18d] 3GPP. Evolved Universal Terrestrial Radio Access (E-UTRA); Physical channels and modulation. Technical Report TR 36.211, Third Generation Partnership Project (3GPP), April 2018.

[3GP18e] 3GPP. Evolved Universal Terrestrial Radio Access (E-UTRA); Physical layer procedures. Technical Report TS. 36.213, Third Generation Partnership Project (3GPP), January 2018.

[3GP18f] 3GPP. Evolved Universal Terrestrial Radio Access (E-UTRA); Radio Resource Control (RRC); Protocol specification. Technical Report TS 36.331, Third Generation Partnership Project (3GPP), January 2018.

[ACC+13] G. Araniti, C. Campolo, M. Condoluci, A. Iera, and A. Molinaro. LTE for vehicular networking: a survey. *IEEE Communications Magazine*, 51(5), pages 148–157, 2013.

[Ahm10] S. Ahmadi. *Mobile WiMAX: A systems approach to understanding IEEE 802.16 m radio access technology.* Academic Press, 2010.

[And13] J. G. Andrews. Seven ways that HetNets are a cellular paradigm shift. *IEEE Communications Magazine*, 51(3), pages 136–144, 2013.

[BAS+05] K. Brueninghaus, D. Astely, T. Salzer, S. Visuri, A. Alexiou, S. Karger, and G. Seraji. Link performance models for system level simulations of broadband radio access systems. In *2005 IEEE 16th International Symposium on Personal, Indoor and Mobile Radio Communications*, pages 2306–2311, 2005.

[BAS10] D. Bültmann, T. Andre, and R. Schoenen. Analysis of 3GPP LTE-Advanced cell spectral efficiency. In *2010 IEEE 21st International Symposium on Personal, Indoor and Mobile Radio Communications*, pages 1876–1881, 2010.

[Ber92] D. P. Bertsekas. Auction algorithms for network flow problems: A tutorial introduction. *Computational optimization and applications*, 1(1), pages 7–66, 1992.

[BPW+18] C. Bockelmann, N. K. Pratas, G. Wunder, S. Saur, M. Navarro, D. Gregoratti, G. Vivier, E. D. Carvalho, Y. Ji, . Stefanović, P. Popovski, Q. Wang, M. Schellmann, E. Kosmatos, P. Demestichas, M. Raceala-Motoc, P. Jung, S. Stanczak, and A. Dekorsy. Towards Massive Connectivity Support for Scalable mMTC Communications in 5G Networks. *IEEE Access*, 6(1), pages 28969–28992, 2018.

[BY12] P. Bao and G. Yu. An interference management strategy for device-to-device underlaying cellular networks with partial location information. In *2012 IEEE 23rd International Symposium on Personal, Indoor and Mobile Radio Communications*, pages 465–470, 2012.

[CLL+10] M. S. Corson, R. Laroia, J. Li, V. Park, T. Richardson, and G. Tsirtsis. Toward proximity-aware internetworking. *IEEE Wireless Communications*, 17(6), pages 26–33, 2010.

[CQH+16] S. Chen, F. Qin, B. Hu, X. Li, and Z. Chen. User-centric ultra-dense networks for 5G: challenges, methodologies, and directions. *IEEE Wireless Communications*, 23(2), pages 78–85, 2016.

[DPSB10] E. Dahlman, S. Parkvall, J. Skold, and P. Beming. *3G evolution: HSPA and LTE for mobile broadband.* Academic Press, 2010.

[DRW+09] K. Doppler, M. Rinne, C. Wijting, C. B. Ribeiro, and K. Hugl. Device-to-device communication as an underlay to LTE-advanced networks. *IEEE Communications Magazine*, 47(12), pages 42–49, 2009.

[dSFM14] J. M. B. da Silva, G. Fodor, and T. F. Maciel. Performance analysis of network-assisted two-hop D2D communications. In *2014 IEEE Globecom Workshops*, pages 1050–1056, 2014.

[EMM+18] S. E. Elayoubi, M. Maternia, J. F. Monserrat, F. Pujol, P. Spapis, V. Frascolla, and D. Sorbara. *Use Cases, Scenarios, and their Impact on the Mobile Network Ecosystem*, chapter 2, pages 15–34. Wiley-Blackwell, 2018.

[ESE13a] Ericsson and ST-Ericsson. 3D-Channel Modeling Extensions and Issues. Technical Report Third Generation Partnership Project (3GPP) TSG-RAN WG1 ♯72, R1-130568, Third Generation Partnership Project (3GPP), January 2013.

[ESE13b] Ericsson and ST-Ericsson. Elevation Angular Modeling and Impact on System Performance. Technical Report Third Generation Partnership Project (3GPP) TSG-RAN WG1 ♯72, R1-130569, Third Generation Partnership Project (3GPP), January 2013.

[ESE13c] Ericsson and ST-Ericsson. Scenarios for 3D-Channel Modeling. Technical Report Third Generation Partnership Project (3GPP) TSG-RAN WG1 ♯72, R1-130567, Third Generation Partnership Project (3GPP), January 2013.

[ETS13] ETSI. Intelligent Transport Systems (ITS); Cooperative ITS (C-ITS); Release 1. Technical Report TR 101 607 V1.1.1, ETSI, May 2013.

[FDM+12] G. Fodor, E. Dahlman, G. Mildh, S. Parkvall, N. Reider, G. Miklós, and Z. Turányi. Design aspects of network assisted device-to-device communications. *IEEE Communications Magazine*, 50(3), pages 170–177, 2012.

[FLYW+13] D. Feng, L. Lu, Y. Yuan-Wu, G. Y. Li, G. Feng, and S. Li. Device-to-Device Communications Underlaying Cellular Networks. *IEEE Transactions on Communications*, 61(8), pages 3541–3551, 2013.

[FR11] G. Fodor and N. Reider. A Distributed Power Control Scheme for Cellular Network Assisted D2D Communications. In *2011 IEEE Global Communications Conference*, pages 1–6, 2011.

[GBCC11] J. Gu, S. J. Bae, B.-G. Choi, and M. Y. Chung. Dynamic power control mechanism for interference coordination of device-to-device communication in cellular networks. In *2011 Third International Conference on Ubiquitous and Future Networks*, pages 71–75, 2011.

[GSA16] A. Gotsis, S. Stefanatos, and A. Alexiou. Ultra Dense Networks: The New Wireless Frontier for Enabling 5G Access. *IEEE Vehicular Technology Magazine*, 11(2), pages 71–78, 2016.

[HHCC13] Huawei, HiSilicon, CATR, and CMCC. WF on evaluation assumptions for SCE physical layer. Technical Report Third Generation Partnership Project (3GPP) TSG-RAN WG1 ♯72, R1-130744, Third Generation Partnership Project (3GPP), January 2013.

[HJS18] B. Han, L. Ji, and H. D. Schotten. Slice as an Evolutionary Service: Genetic Optimization for Inter-Slice Resource Management in 5G Networks. *IEEE Access*, 6(1), pages 33137–33147, 2018.

[HMCK05] L. Hanzo, M. Münster, B. Choi, and T. Keller. *OFDM and MC-CDMA for broadband multi-user communications, WLANs and broadcasting*. John Wiley & Sons, 2005.

[HS13] Z. Hanzaz and H. D. Schotten. Impact of L2S interface on system level evaluation for LTE system. In *2013 IEEE 11th Malaysia International Conference on Communications*, pages 456–461, 2013.

[HS17] B. Han and H. D. Schotten. Grouping-Based Random Access Collision Control for Massive Machine-Type Communication. In *2017 IEEE Global Communications Conference*, pages 1–7, 2017.

[HT07] H. Holma and A. Toskala. *WCDMA for UMTS: HSPA Evolution and LTE*. John Wiley & Sons, 2007.

[HT09] H. Holma and A. Toskala. *LTE for UMTS: OFDMA and SC-FDMA based radio access*. John Wiley & Sons, 2009.

[IEE10] IEEE. Amendment 6: Wireless Access in Vehicular Environments. Technical Report IEEE Std 802.11p, IEEE, July 2010.

[IR08a] ITU-R. Guidelines for evaluation of radio interface technologies for IMT-Advanced. Technical Report M.2135, ITU Radiocommunication Sector (ITU-R), 2008.

[IR08b] ITU-R. Requirements related to Technical performance for IMT-Advanced Radio Interface(s). Technical Report M.2134, ITU Radiocommunication Sector (ITU-R), 2008.

[IR14] ITU-R. Future technology trends of terrestrial IMT systems. Technical Report M.2320, ITU Radiocommunication Sector (ITU-R), November 2014.

[IR15] ITU-R. IMT vision–framework and overall objectives of the future development of IMT for 2020 and beyond. Technical Report M. 2083-0, ITU Radiocommunication Sector (ITU-R), 2015.

[IWR10] J. C. Ikuno, M. Wrulich, and M. Rupp. System level simulation of LTE networks. In *2010 IEEE 71st Vehicular Technology Conference*, 2010.

[Jai08] R. Jain. WiMAX system evaluation methodology v2. 1. In *WiMAX Forum*, 2008.

[JDWS18] L. Ji, W. Donglin, A. Weinand, and H. D. Schotten. Network-assisted Two-hop Vehicle-to-Everything Communication on Highway. In *2018 IEEE 87th Vehicular Technology Conference*, pages 1–5, 2018.

[JF12] M. Jar and G. Fettweis. Throughput maximization for LTE uplink via resource allocation. In *The 9th International Symposium on Wireless Communication Systems*, pages 146–150, 2012.

[JHLS17] L. Ji, B. Han, M. Liu, and H. D. Schotten. Applying Device-to-Device Communication to Enhance IoT Services. *IEEE Communications Standards Magazine*, 1(2), pages 85–91, 2017.

[JKK+14] L. Ji, A. Klein, N. Kuruvatti, R. Sattiraju, and H. D. Schotten. Dynamic Context-Aware Optimization of D2D Communications. In *2014 IEEE 79th Vehicular Technology Conference*, pages 1–5, 2014.

[JKKS14] L. Ji, A. Klein, N. Kuruvatti, and H. D. Schotten. System capacity optimization algorithm for D2D underlay operation. In *2014 IEEE International Conference on Communications*, pages 85–90, 2014.

[JLS17] L. Ji, M. Liu, and H. D. Schotten. Context-aware cluster based device-to-device communication to serve machine type communications. In *2017 IEEE International Conference on Communications*, pages 1011–1017, 2017.

[JLWS17] L. Ji, M. Liu, A. Weinand, and H. D. Schotten. Direct vehicle-to-vehicle communication with infrastructure assistance in 5G network. In *2017 16th Annual Mediterranean Ad Hoc Networking Workshop*, pages 1–5, 2017.

[JT13] L. Ji and W. G. Teich. An Improved Chip Detector for OFDM-IDMA over Fast Time-Variant Channels. In *The 10th International Symposium on Wireless Communication Systems*, pages 1–5, 2013.

[JWHS17] L. Ji, A. Weinand, B. Han, and H. D. Schotten. Feasibility Study of Enabling V2X Communications by LTE-Uu Radio Interface. In *IEEE/CIC International Conference on Communications in China*, pages 1–5, 2017.

[JWHS18a] L. Ji, A. Weinand, B. Han, and H. D. Schotten. Applying Multiradio Access Technologies for Reliability Enhancement in Vehicle-to-Everything Communication. *IEEE Access*, 6(1), pages 23079–23094, 2018.

[JWHS18b] L. Ji, A. Weinand, B. Han, and H. D. Schotten. Multi-RATs support to improve V2X communication. In *2018 IEEE Wireless Communications and Networking Conference*, pages 1–5, 2018.

[JWKS17] L. Ji, A. Weinand, M. Karrenbauer, and H. D. Schotten. Radio Link Enabler for Context-Aware D2D Communication in Reuse Mode. In *2017 IEEE 85th Vehicular Technology Conference*, pages 1–7, 2017.

[KKJ+15] N. P. Kuruvatti, A. Klein, L. Ji, C. Zhou, O. Bulakci, J. Eichinger, R. Sattiraju, and H. D. Schotten. Robustness of Location Based D2D Resource Allocation against Positioning Errors. In *2015 IEEE 81st Vehicular Technology Conference*, pages 1–6, 2015.

[KRSS14] A. Klein, A. Rauch, R. R. Sattiraju, and H. D. Schotten. Achievable Performance Gains Using Movement Prediction and Advanced 3D System Modeling. In *2014 IEEE 79th Vehicular Technology Conference*, pages 1–5, 2014.

[KWJS18] M. Karrenbauer, A. Weinand, L. Ji, and H. D. Schotten. Network Slicing in Local Non-Cellular Wireless Networks: A MC-CDMA-based Approach. In *The 15th International Symposium on Wireless Communication Systems*, pages 1–6, 2018.

[LAGR14] X. Lin, J. G. Andrews, A. Ghosh, and R. Ratasuk. An overview of 3GPP device-to-device proximity services. *IEEE Communications Magazine*, 52(4), pages 40–48, 2014.

[LCL11] S. Lien, K. Chen, and Y. Lin. Toward ubiquitous massive accesses in 3GPP machine-to-machine communications. *IEEE Communications Magazine*, 49(4), pages 66–74, 2011.

[Lin06] J. Lindner. *Informationsübertragung: Grundlagen der Kommunikationstechnik.* Springer-Verlag, 2006.

[LJ12] A. Lozano and N. Jindal. Are yesterday-s information-theoretic fading models and performance metrics adequate for the analysis of today's wireless systems? *IEEE Communications Magazine*, 50(11), pages 210–217, 2012.

[LTRa+18] J. Lee, E. Tejedor, K. Ranta-aho, H. Wang, K. T. Lee, E. Semaan, E. Mohyeldin, J. Song, C. Bergljung, and S. Jung. Spectrum for 5G: Global Status, Challenges, and Enabling Technologies. *IEEE Communications Magazine*, 56(3), pages 12–18, 2018.

[Mal12] D. Malladi. Heterogeneous Networks in 3G and 4G. In *IEEE Communication Theory Workshop*, 2012. `http://ctw2012.ieee-ctw.org/HetNets3Gand4GIEEECTW2012.pdf`.

[MBSW11] M. Mühleisen, D. Bültmann, R. Schoenen, and B. Walke. Analytical evaluation of an IMT-advanced compliant LTE system level simulator. In *European Wireless Conference 2011-Sustainable Wireless Technologies*, pages 1–5, 2011.

[MET13a] METIS. Scenarios, requirements and KPIs for 5G mobile and wireless system. Technical Report Deliverable 1.1, METIS Project, April 2013.

[MET13b] METIS. Simulation guidelines. Technical Report Deliverable 6.1, METIS Project, October 2013.

[MET14a] METIS. Initial report on horizontal topics, first results and 5G system concept. Technical Report Deliverable 6.2, METIS Project, March 2014.

[MET14b] METIS. Intermediate system evaluation results. Technical Report Deliverable 6.3, METIS Project, August 2014.

[MET16] METISII. Performance evaluation framework. Technical Report Deliverable 2.1, METIS-II Project, January 2016.

[MET17a] METISII. Final air interface harmonization and user plane design. Technical Report Deliverable 4.2, METIS-II Project, April 2017.

[MET17b] METISII. Final asynchronous control functions and overall control plane design. Technical Report Deliverable 6.2, METIS-II Project, April 2017.

[MET17c] METISII. Final Considerations on Synchronous Control Functions and Agile Resource Management for 5G. Technical Report Deliverable 5.2, METIS-II Project, March 2017.

[MFMSJ16] J. F. Monserrat, M. Fallgren, D. Martin-Sacristan, and L. Ji. *Simulation methodology*, chapter 14, pages 381–397. Cambridge University Press, 2016.

[MKH11] T. Mangel, O. Klemp, and H. Hartenstein. 5.9 GHz inter-vehicle communication at intersections: a validated non-line-of-sight path-loss and fading model. *EURASIP Journal on Wireless Communications and Networking*, 2011(1), pages 182–192, 2011.

[MKM+10] A. Maltsev, A. Khoryaev, R. Maslennikov, M. Shilov, M. Bovykin, G. Morozov, A. Chervyakov, A. Pudeyev, V. Sergeyev, and A. Davydov. Analysis of IEEE 802.16 m and 3GPP LTE Release 10 technologies by Russian Evaluation Group for IMT-Advanced. In *2010 IEEE International Congress on Ultra Modern Telecommunications and Control Systems and Workshops*, pages 901–908, 2010.

[MKSS11] C. Mannweiler, A. Klein, J. Schneider, and H. D. Schotten. Context-based user grouping for multi-casting in heterogeneous radio networks. *Advances in Radio Science*, 9(1), pages 187–193, 2011.

[MMG17] R. Molina-Masegosa and J. Gozalvez. LTE-V for Sidelink 5G V2X Vehicular Communications: A New 5G Technology for Short-Range Vehicle-to-Everything Communications. *IEEE Vehicular Technology Magazine*, 12(4), pages 30–39, 2017.

[MRA17] A. Memmi, Z. Rezki, and M. S. Alouini. Power Control for D2D Underlay Cellular Networks With Channel Uncertainty. *IEEE Transactions on Wireless Communications*, 16(2), pages 1330–1343, 2017.

[MSC+13] C. Mannweiler, J. Schneider, P. Chakraborty, A. Klein, and H. D. Schotten. A distributed broker system enabling coordinated access schemes in autonomous wireless networks. In *2013 IEEE 9th International Wireless Communications and Mobile Computing Conference*, pages 59–64, 2013.

[MSKS11] C. Mannweiler, J. Schneider, A. Klein, and H. D. Schotten. From context to context-awareness: model-based user classification for efficient multicasting. In *International Conference on Knowledge-Based and Intelligent Information and Engineering Systems*, pages 146–154, 2011.

[MSMOC11] D. Martín-Sacristán, J. Monserrat, V. Osa, and J. Cabrejas. LTE-advanced system level simulation platform for IMT-advanced evaluation. In *Waves*, pages 15–24, 2011.

[MVSAT+18] F. J. Martín-Vega, B. Soret, M. C. Aguayo-Torres, I. Z. Kovács, and G. Gómez. Geolocation-Based Access for Vehicular Communications: Analysis and Optimization via Stochastic Geometry. *IEEE Transactions on Vehicular Technology*, 67(4), pages 3069–3084, 2018.

[NGM15] NGMN. 5G White Paper By NGMN Alliance. Technical report, NGMN, February 2015.

[NN13] N. S. Networks and Nokia. Detailed 3D Channel Model. Technical Report Third Generation Partnership Project (3GPP) TSG-RAN WG1 ♯72, R1-130500, Third Generation Partnership Project (3GPP), January 2013.

[OBB+14] A. Osseiran, F. Boccardi, V. Braun, K. Kusume, P. Marsch, M. Maternia, O. Queseth, M. Schellmann, H. D. Schotten, H. Taoka, H. Tullberg, M. A. Uusitalo, B. Timus, and M. Fallgren. Scenarios for 5G mobile and wireless communications: the vision of the METIS project. *IEEE Communications Magazine*, 52(5), pages 26–35, 2014.

[OBH+13] A. Osseiran, V. Braun, T. Hidekazu, P. Marsch, H. D. Schotten, H. Tullberg, M. A. Uusitalo, and M. Schellman. The Foundation of the Mobile and Wireless Communications System for 2020 and Beyond: Challenges, Enablers and Technology Solutions. In *2013 IEEE 77th Vehicular Technology Conference*, pages 1–5, 2013.

[OMM16] A. Osseiran, J. F. Monserrat, and P. Marsch. *5G Mobile and Wireless Communications Technology.* Cambridge University Press, 2016.

[Oss11] A. Osseiran. *Mobile and wireless communications for IMT-advanced and beyond.* John Wiley & Sons, 2011.

[PP13] N. Pratas and P. Popovski. Low-Rate Machine-Type Communication via Wireless Device-to-Device (D2D) Links. 2013. `https://arxiv.org/pdf/1305.6783.pdf`.

[PP14] N. K. Pratas and P. Popovski. Underlay of low-rate machine-type D2D links on downlink cellular links. In *2014 IEEE International Conference on Communications Workshops*, pages 423–428, 2014.

[PR91] M. Padberg and G. Rinaldi. A branch-and-cut algorithm for the resolution of large-scale symmetric traveling salesman problems. *SIAM review*, 33(1), pages 60–100, 1991.

[RJKS15] A. Rauch, L. Ji, A. Klein, and H. D. Schotten. Fast algorithm for radio propagation modeling in realistic 3-D urban environment. *Advances in Radio Science*, 13(1), pages 169–173, 2015.

[SBT11] S. Sesia, M. Baker, and I. Toufik. *LTE - the UMTS long term evolution: from theory to practice.* John Wiley & Sons, 2011.

[SJD+18] S. Shubhranshu, L. Ji, C. Daniel, G.-R. David, M. N. H., P. Nuno, M. Tomasz, and G. M. Carmela. *D2D and V2X Communications*, chapter 14, pages 409–449. Wiley-Blackwell, 2018.

[SJMS13] J. Schneider, L. Ji, C. Mannweiler, and H. D. Schotten. Context-based Cognitive Radio for LTE-Advanced Networks. In *Proceedings of the 18th ITG Fachtagung Mobilkommunikation*, pages 42–46, 2013.

[SKC10] D. Singhal, M. Kunapareddy, and V. Chetlapalli. LTE-Advanced: latency analysis for IMT-A evaluation. *White paper, Tech Mahindra Limited*, 2010.

[SKC+11] D. Singhal, M. Kunapareddy, V. Chetlapalli, V. B. James, and N. Akhtar. LTE-advanced: handover interruption time analysis for IMT-A evaluation. In *2011 IEEE International Conference on Signal Processing, Communication, Computing and Networking Technologies*, pages 81–85, 2011.

[SKCH11] J. Seppälä, T. Koskela, T. Chen, and S. Hakola. Network controlled Device-to-Device (D2D) and cluster multicast concept for LTE and LTE-A networks. In *2011 IEEE Wireless Communications and Networking Conference*, pages 986–991, 2011.

[SKJ+15] R. Sattiraju, A. Klein, L. Ji, C. Zhou, O. Bulakci, J. Eichinger, N. P. Kuruvatti, and H. D. Schotten. Virtual Cell Sectoring for Enhancing Resource Allocation and Reuse in Network Controlled D2D Communication. In *2015 IEEE 81st Vehicular Technology Conference*, pages 1–6, 2015.

[SKMS12] J. Schneider, A. Klein, C. Mannweiler, and H. D. Schotten. A context management system for a cost-efficient smart home platform. *Advances in Radio Science*, 10(1), pages 135–139, 2012.

[SPMS13] J. Santa, F. Pereñíguez, A. Moragón, and A. F. Skarmeta. Vehicle-to-infrastructure messaging proposal based on CAM/DENM specifications. In *2013 IFIP Wireless Days*, pages 1–7, 2013.

[SPMS14] J. Santa, F. Pereñíguez, A. Moragón, and A. F. Skarmeta. Experimental evaluation of CAM and DENM messaging services in vehicular communications. *Transportation Research Part C: Emerging Technologies*, 46(1), pages 98–120, 2014.

[SSPR16] I. Safiulin, S. Schwarz, T. Philosof, and M. Rupp. Latency and Resource Utilization Analysis for V2X Communication over LTE MBSFN Transmission. In *20th International ITG Workshop on Smart Antennas*, pages 1–6, 2016.

[STB10] R. Schoenen, C. Teijeiro, and D. Bultmann. System level performance evaluation of LTE with MIMO and relays in reuse-1 IMT-Advanced scenarios. In *2010 IEEE 6th International Conference on Wireless Communications Networking and Mobile Computing*, pages 1–5, 2010.

[Stu96] G. L. Stueber. *Principles of mobile communication*, volume 2. Springer, 1996.

[SZJ+09] R. Srinivasan, J. Zhuang, L. Jalloul, R. Novak, and J. Park. IEEE 802.16 m evaluation methodology document (EMD). Technical Report IEEE 802.16m-08/004r5, IEEE, January 2009.

[TJ14] W. G. Teich and L. Ji. Variance Transfer Chart Analysis of an OFDM-IDMA System with Carrier Frequency Offset. In *2014 IEEE 80th Vehicular Technology Conference*, pages 1–6, 2014.

[TV05] D. Tse and P. Viswanath. *Fundamentals of wireless communication*. Cambridge University Press, 2005.

[Vin12] A. Vinel. 3GPP LTE Versus IEEE 802.11p/WAVE: Which Technology is Able to Support Cooperative Vehicular Safety Applications? *IEEE Wireless Communications Letters*, 1(2), pages 125–128, 2012.

[Wal99] B. H. Walke. *Mobile radio networks*, volume 2. John Wiley & Sons, 1999.

[WCC+11] B. Wang, L. Chen, X. Chen, X. Zhang, and D. Yang. Resource Allocation Optimization for Device-to-Device Communication Underlaying Cellular Networks. In *2011 IEEE 73rd Vehicular Technology Conference*, pages 1–6, 2011.

[WKJS17] A. Weinand, M. Karrenbauer, L. Ji, and H. D. Schotten. Physical Layer Authentication for Mission Critical Machine Type Communication Using Gaussian Mixture Model Based Clustering. In *2017 IEEE 85th Vehicular Technology Conference*, pages 1–5, 2017.

[WWFD18] Q. Wei, L. Wang, Z. Feng, and Z. Ding. Wireless Resource Management in LTE-U Driven Heterogeneous V2X Communication Networks. *IEEE Transactions on Vehicular Technology*, 67(8), pages 7508–7522, 2018.

[WWLS10] Y. Wang, T. Wang, L. Ling, and C. Shi. On the evaluation of IMT-advanced candidates: Methodologies, current work and way forward. In *2010 IEEE 6th International Conference on Wireless Communications Networking and Mobile Computing*, pages 1–4, 2010.

[YDRT11] C. H. Yu, K. Doppler, C. B. Ribeiro, and O. Tirkkonen. Resource Sharing Optimization for Device-to-Device Communication Underlaying Cellular Networks. *IEEE Transactions on Wireless Communications*, 10(8), pages 2752–2763, 2011.

[YTDR09] C. H. Yu, O. Tirkkonen, K. Doppler, and C. Ribeiro. On the Performance of Device-to-Device Underlay Communication with Simple Power Control. In *IEEE 69th Vehicular Technology Conference*, pages 1–5, 2009.

Curriculum Vitae

Personal Details

Name:	Ji Lianghai
Place of birth:	Shandong, China

Professional Experience

03/2019 - Present	Nokia Bell Labs Standardization Research Lab Device\modem standards research expert
11/2012 - 10/2018	Technische Universität Kaiserslautern Wireless Communications and Navigation research group Research and teaching associate

Education

11/2012 - 10/2019	Technische Universität Kaiserslautern PhD thesis: "Design and Application of D2D and V2X Communications in the 5G Radio Access Network"
04/2010 - 07/2012	University of Ulm Master degree in Communications Technology Thesis: "Implementation and Analysis of the Wireless Communication System OFDM-IDMA"
09/2006 - 06/2010	Shandong University Bachelor degree in Electronic Information Science and Technology Thesis: "Analysis Based on the Time-Space Coding of Alamouti"